a collector's guide to
minerals and gemstones

a collector's guide to
MINERALS and GEMSTONES

HELLMUTH BOEGEL

edited and revised by
JOHN SINKANKAS

with 154 illustrations
in colour by
Claus Caspari

THAMES AND HUDSON

Translated from the German
Knaurs Mineralienbuch by
Eva Fejer and Patricia Walker, F.G.A.

© 1968 by Droemersche Verlagsanstalt Munich/Zurich
© This edition 1971 by Thames and Hudson Limited, London

Filmset in Great Britain by Keyspools Ltd, Golborne, Lancs.
Printed in the German Democratic Republic
for Thames and Hudson by Interdruck, Leipzig

Clothbound 0 500 01066 8
Paperbound 0 500 27012 0

Contents

Preface

Soon after Hellmuth Boegel's immensely popular handbook of minerals appeared in Germany I obtained a copy for my personal library and was immediately impressed by its factual, concise text, happily combined with a series of splendid color plates prepared especially for the work. My high regard for this book must have made itself felt to the present publishers for within several months they asked me to render an opinion as to its suitability for translation and publication into English.

My enthusiastic response to this inquiry promptly led to a further request that I undertake revision of the translation to make it more useful to its expected audience. To this I agreed.

In revising the text I was mindful of two important object-ives of the author: to provide a fuller treatment of mineral origin and mineral deposit formation than is usually accorded in similar works meant for the amateur, and to provide details of a historical and economic nature on many species which are important in some way in the arts, sciences, or industry.

The first objective is laudable because the genesis of minerals and their concentration in deposits is knowledge directly beneficial to anyone studying mineralogy. From such knowl-edge comes the important realization that minerals are not scattered randomly but form limited groupings or associations, which, when recognized, greatly reduce the problems of identification and provide valuable guidance when examining deposits during field trips. For these reasons the discussions of the original text are carried over more or less completely into this version.

In respect to the second objective, the uses to which minerals are put is cultural rather than scientific in nature, and therefore sections containing such discussions have been considerably modified to make room for additional details on the minerals themselves as well as to strengthen remarks on distinctive features useful in identification of individual species, to make room for much more data on important localities, and to furnish additional detail on mineral specimens obtained therefrom. In this connection, judicious deletions of certain obscure European occurrences and economically important but specimen-poor deposits provided further space for inclusion of numerous modern and classical localities omitted from the original text. The emphasis understandably placed by the author upon German and other Continental occurrences is now more evenly allocated among world occurrences as a whole, without, however, neglecting the truly important European localities.

As an amateur handbook, this one makes no pretense at being a textbook. Only enough information is given in the preliminary theoretical portion to acquaint the reader with the essentials of crystallography, chemistry, and crystal structure. These provide a firm basis for discussion of resulting physical properties and lend meaning and usefulness to the chemical and physical data placed under each species in the descriptive portion. The changes made in such parts of the text are additive, mainly by providing fuller detail aimed at making distinctive features of individual species of greater value to the reader, and as mentioned before, providing ideas as to size and quality of specimens. The time-honored method of indicating increasing quality by the use of an increasing number of exclamation points within parentheses, e.g., (!), (!!), etc., is restored and should quickly invite the reader's attention to sources of superb specimens.

In the discussion of crystallography at the beginning of the text, crystal systems and associated conventions and terminology were changed to reflect current usage in the United States in which the largest reader population is expected to exist. This consideration also dictated the choice of nomenclature and terminology elsewhere, sometimes requiring careful checking against the German text to insure that the word and spirit in this version are truly those intended in the original. As an outgrowth of this study a number of errors in the original were detected and corrected.

My admiration of the color plates is based on my personal experience in making similar water color illustrations for one of my own books.* It is far from a simple task because the

* *Gemstones of North America.*

artist must know minerals in order to avoid those lamentable errors in crystal forms and colors which are instantly apparent to the expert and, at best, seriously mislead the inexpert. The superb color illustrations were prepared over a period of several years by Claus Caspari from real specimens. Most are in natural size, some slightly reduced, and a few enlarged, as the diamond crystal of Plate 1, depicted in both natural and greatly magnified size to show clearly the interesting and characteristic growth figures upon its faces. Caspari's renditions are all the more remarkable because he is an ardent amateur mineral collector and is therefore completely conversant with correct rendering of crystals, lusters, and hues, adjusting these factors in a manner no color camera can do to bring before the eye the truest representations possible.

Finally it should be mentioned that the works of reference listed at the end of the text differ materially from those of the original to include standard texts in the English language.

JOHN SINKANKAS

San Diego, California
June, 1970

Introduction

Go to any river or stream where the water has exposed extensive sand and gravel banks. You will see inconspicuous boulders lying about, most of them rounded, but some more oval, flat, or spherical. They all seem to be equally grey and dull-looking. But are they? After a closer look one appears to be slightly reddish, another yellowish-grey, a third one mottled. There is a greyish-green boulder here with faint, pale, reddish spots. Break it open, and you will see a tangled mass of fine, blackish-green needles with shiny red grains between them. The freshly fractured surface of another boulder reveals silvery flakes between glassy, dull, yellowish grains. A third one will be very disappointing when broken. The outside and the interior are both grey and uniform.

Take the three pieces with you and show them to someone who knows about stones, to a collector perhaps, to a stone-mason, or to a mineralogist. You will hear some familiar names but also some weird-sounding ones. You will be told that the uninteresting-looking grey piece is limestone. There is nothing to be seen on it, is there? But if you are offered a magnifying glass to look at it with, the dense mass will be resolved into tiny grains, and minute shiny surfaces become visible. You will be told that these are fragments of calcite crystals. Three minerals will be mentioned to you for the second piece: the flakes are mica, the glassy grains quartz, and the yellowish grains feldspar. The whole specimen is called granite. Limestone and granite you will already have heard about. Your third piece, however, will be described to you as amphibolite, a totally strange, weird name to you. The green needles are hornblende, the red grains are nothing else but garnets, the same 'garnet' which, when cut and mounted, can be seen in any jeweler's window.

ROCKS, MINERALS, CRYSTALS

What does the word 'mineral' really mean? Unfortunately the answer to this is not simple and straightforward. 'Mineral', 'rock', and 'stone' are used indiscriminately as equivalents, but this is wrong. It is true that all rocks consist of minerals, but a mineral is not always a rock. On the other hand, gold and silver, found pure or in the 'native' state, are indeed minerals – no one would ever refer to them as stones.

A short, but of course somewhat dull definition of the terms 'mineral', 'rock', and 'crystal' cannot be avoided at this stage. A *rock* is an aggregate of mineral grains usually of the order of 0.1–1 mm. in size (with variations in both directions of course). A rock must have been formed by a natural process; concrete and bricks are not to be included here. Such an aggregate of mineral grains must also possess widespread distribution. A single, chance association of minerals or occasional occurrence in small quantities does not constitute a rock. Sometimes a rock consists of only one kind of mineral, but usually of two or three minerals, and occasionally of a few more. Limestone or marble are good examples illustrating the first case, consisting (apart from impurities) only of the mineral calcite. Granite is the best-known example of a rock consisting of several minerals, of which feldspar, quartz, and mica are the main constituents. Quartzite consists of quartz only, mica schist of quartz and mica, and eclogite of garnet and augite. There are also rocks which consist of fragments of other rocks; these are the con-glomerates and breccias.

In contrast to rocks, which are mixtures, minerals have *definite compositions*. Like rocks, minerals too must be formed naturally and they must be constituents of the Earth's crust. These three conditions must be explained in a little more detail. A mineral may consist of a single chemical element, such as sulphur or gold, but more often it consists of a chemical compound. The elements sodium and chlorine in combination form halite (rock salt). Several minerals consist of a large number of elements and are very complicated compounds, but any one part of a mineral must have the same composition and proper-ties as any other part; the composition and properties must be constant all over a given mineral. Or, to illustrate this point, if a mineral is powdered, then each particle of the powder must still have the properties of the original sample. The other two conditions are easily understood. Artificial products such as glass, synthetic chemical compounds, or artificially grown crystals do not count as minerals.

The reader will have noticed that up to now crystals have not been mentioned, in spite of the fact that with their regular forms they are probably the most striking representatives of the mineral kingdom. The term 'crystal' nowadays has a different meaning from that originally given to it. Crystals are uniform bodies whose atoms possess a regular geometric arrangement. This regular array may result in a regular external form, but this is not essential. If, for example, a rock crystal is broken up into irregular fragments, the external crystalline form will have been destroyed but not the regularity of the internal structure and the fragments will still be crystalline. The terms 'crystal' and 'crystalline' are also used to describe artificial substances, provided that they have a regular internal structure. The terms 'mineral' and 'crystal' are therefore not always equivalent. There are even some minerals, such as opal, which do not have a strictly regular internal structure.

MINERALS AND MAN

Man's first encounter with minerals was in Paleolithic times, when he recognized that some of the many stones around him were particularly suitable for the carving of implements and tools. As long as half a million years ago flint was the favorite stone for making the famous hand axes. Flints had to be identified and located, making ancient man the first mineralogist, and it was no longer possible to lead an everyday life without a knowledge of mineral raw materials. The manufacture of stone implements reached its peak in Neolithic times, and there is evidence of at least one flint mine dating from this period. Finely worked and polished tools and weapons made of flint, nephrite, and other tough stones are found all over the world, and their production clearly ranked with that of food supplies.

Copper and gold were the first known metals. Gold occurs openly in nature. Its color makes it conspicuous and it can be easily worked, but it is too soft for making tools. It was used in jewelry, as it still is today. The first metal tools were made of copper, occasionally found native, but pure copper is also too soft. Only by the addition of tin did man obtain a usable alloy, bronze. Tin only occurs in important quantity in cassiterite (tin and oxygen), and never as the pure metal. In order to obtain copper in sufficient quantities copper ores had to be found and smelted. Remains of Bronze Age copper mines can be found, for instance, in various parts of the Austrian Tirol and Salzburg. Much tin ore came from Britain so that brisk trading in minerals was already going on then. This trading also included rock salt, and the search for this valued mineral led early man to burrow deep into the interior of mountains. An important prehistoric culture, the Hallstatt culture, derives its name from the salt-mining district of Hallstatt in Upper Austria, where many prehistoric finds have been made. In the course of present-day salt-mining, still flourishing there, traces of ancient workings have been repeatedly encountered, revealing not only tools, clothing, and remnants of food, but remains of Man himself, entombed and preserved in salt, the very mineral he endeavored to obtain.

Man's need for ornaments seems to be as old as his knowledge of toolmaking. Thus we know of pierced flints used as pendants dating from the early Paleolithic. Dating from Neolithic times, some 100,000 years ago, beads made of limestone and quartz were found, and when man began to decorate cave walls with crude drawings and paintings, probably for reasons of cult, he had the choice of three colors: black, red and yellowish-brown. Two of these, red and yellowish-brown, were supplied by the mineral kingdom: minium, cinnabar, and limonite (ochre). The black color was soot.

By Egyptian, Greek and Roman times numerous metals and their ores were known to man, as well as many gemstones, and the first written accounts of minerals date from that period. These do not merely tell us about the occurrence and mining of minerals but also of the many magic and medicinal qualities attributed to various minerals, ideas which persisted to medieval times. Thus we learn that the Greeks used to wear purple rock crystal as amulets, or that in pulverized form it was taken as medicine as a protection against drunkenness – the name amethyst in fact means 'not drunken'. Clear transparent rock crystal was considered by the Greeks to be ice, frozen at such low temperatures that it was no longer possible to melt it. It was supposed to have many magic properties, and as a general protective amulet it prevented hemorrhages, and cured dropsy and even toothache. In medieval times it was thought that rock crystal protected against thirst, and magnificent vases and jugs were cut from many a large rock crystal; since they were certainly not mere ornaments, the rock crystal in this form was meant to fulfil a protective role. Green serpentine was used against snake bites, and poisons in general. Topaz, powdered in wine, chased away depression – a result which wine alone, without the addition of topaz, could easily achieve! Albertus Magnus, a famous alchemist of the thirteenth century, tells us that opal wrapped in bay leaf made its wearer invisible, and therefore called it the 'patron of thieves'.

One should not dismiss these fanciful opinions with a smile, because even in present times we are by no means free from superstition or pure sentiment where minerals are concerned. Opal is still thought to bring bad luck, and similar stories are told about certain famous gemstones, particularly diamonds. Minerals play their part in astrology as 'birthstones', and there is not a single gemstone lover or mineral collector who does not find certain stones more attractive than others.

Part One GENERAL MINERALOGY

1 The structure and composition of minerals

CHEMICAL COMPOSITION

As already mentioned, a mineral consists either of a single chemical element, or it is a chemical compound consisting of two or more elements. In compounds the atoms, the smallest parts of an element, are in numerically constant proportions to one another. They are not mere mixtures of various elements in random proportions. For example, rock salt contains equal numbers of chlorine and sodium atoms combined. If salt is dissolved in water the sodium and chlorine are separated and move about freely, not as atoms but as charged particles called *ions*. Each chlorine ion bears a negative charge and each sodium ion a positive one. In a salt crystal, however, the particles are in a regular geometric arrangement known as a crystal lattice (see p. 27 and fig. 9). The over-all ratio 1 : 1 chlorine to sodium remains, but each element atom is surrounded by six of the other kind. Similarly, pyrite consists of one part of iron to two parts of sulphur, magnesite of one part magnesium to one part carbon to three parts oxygen, and so on, each with its own crystal arrangement.

Abbreviations of each element were introduced, so that minerals as well as artificial compounds could be expressed as *formulas*. Gold (Au), carbon (C), rock salt (NaCl), pyrite (FeS_2), magnesite ($MgCO_3$), potash feldspar ($KAlSi_3O_8$), are mineral examples, whilst hydrochloric acid (HCl), sulphuric acid (H_2SO_4), potassium hydroxide (KOH), are examples of artificial compounds. Water consists of two parts hydrogen and one part oxygen (H_2O). Of the hundred or so known elements the mineralogically most important are listed alphabetically in Table 1, together with their abbreviations. Elements fall into three main groups: gases (Cl, O, H, etc.), metals (Fe, Cu, Au, Na, Al etc.), and non-metals (C, S, P, Si, etc.). A few semimetals

(As, Sb etc.) occupy an intermediate position between the last two. Properties of compounds are totally different from the properties of their constituent elements. Poisonous chlorine gas, for instance, combines with sodium metal to form rock salt, indispensable for human existence.

Viewed diagrammatically, each normal chemical compound consists of one positively and one negatively charged ion or group of ions: a positive *cation* $(+)$, and a negative *anion* $(-)$ or acid radical. These charges must always balance: Na^+Cl^-, $Fe^{2+}S_2^-$, $Mg^{2+}[CO_3]^{2-}$, $K^+[AlSi_3O_8]^-$, H^+Cl^-, $H_2^+[SO_4]^{2-}$, $K^+[OH]^-$, etc.

TABLE 1 The most important chemical elements, in alphabetical order of symbols.

(G = gas, N = non-metal, SM = semi-metal, M = metal.)

Ag	silver		F	fluorine	G	Pt	platinum	M
Al	aluminium	M	Fe	iron	M	Ra	radium	M
As	arsenic	SM	H	hydrogen	G	S	sulphur	N
An	gold	M	Hg	mercury	M	Sb	antimony	SM
B	boron	N	K	potassium	M	Si	silicon	M
Ba	barium	M	Li	lithium	M	Sn	tin	M
Be	beryllium	M	Mg	magnesium	M	Sr	strontium	M
C	carbon	N	Mn	manganese	M	Ta	tantalum	M
Ca	calcium	M	Mo	molybdenum	M	Ti	titanium	M
Cd	cadmium	M	N	nitrogen	G	U	uranium	M
Ce	cerium	M	Na	sodium	M	V	vanadium	M
Cl	chlorine	G	Nb	niobium	M	W	tungsten	M
Co	cobalt	M	O	oxygen	G	Y	yttrium	M
Cr	chromium	M	P	phosphorus	N	Zn	zinc	M
Cu	copper	M	Pb	lead	M	Zr	zirconium	M

The number of charges due to each element is known as its valency. Thus sodium is univalent positive, chlorine univalent negative, aluminum trivalent positive, nitrogen trivalent negative, carbon and silicon are tetravalent positive, oxygen bivalent negative, etc. Many elements have more than one valency. Iron can be bi- or trivalent, Fe^{2+} and Fe^{3+}, copper uni- or bivalent, Cu^+ or Cu^{2+}.

Because minerals are natural products, they obey these rules only within certain limits. Rarely are minerals entirely free from impurities, so that massive types and even well-formed crystals may contain grains of other minerals as inclusions; this can lead to errors when determining their chemical composition. Furthermore the composition of many minerals can also vary to some extent without affecting their external properties. In magnesite, for instance, Fe^{2+} may substitute for Mg^{2+} up to an Mg : Fe ratio of 9 : 1, without causing obvious changes in the mineral. Sphalerite, ZnS, practically always contains some iron, but only more than 10 per cent iron causes the colour to become dark brown and eventually black, and the luster somewhat metallic. Such an iron-rich form is not a separate species but remains a variety (p. 29). Sphalerite never contains more than about 20 per cent of iron, whereas magnesite may contain any amount of iron until the magnesium is completely replaced and the formula becomes $FeCO_3$. This now becomes a new mineral, siderite (p. 155).

Silicate minerals form a large part of the mineral kingdom partly because tetravalent silicon may be replaced by trivalent Al, which alters the valency of the anion and allows great variation in composition. Plagioclase feldspars provide a good example of this (compare p. 246 and Table 11):

$$Na^+[AlSi_3O_8]^- \rightleftharpoons Ca^{2+}[Al_2Si_2O_8]^{2-}$$

Univalent Na ions are progressively replaced by bivalent Ca, and equal amounts of Al simultaneously replace Si, maintaining the electrical charges in balance. *End members* of such continuous *series* are distinct, whereas intermediate members are recognized only after very detailed examination. This replacement of certain elements by others without affecting the basic structure is known as *isomorphous substitution* or *diadochy*.

In isomorphous series the extent of substitution may be limited as in sphalerite, or unlimited as in the plagioclases or in magnesite-siderite.

Even the law of balancing charges is sometimes disobeyed by Nature. Thus pyrrhotite FeS, for example, nearly always contains a deficit of iron, its formula being nearer to Fe_7S_8 than to FeS.

PLATE 1   *Elements*
Above left Wire silver, × 1½, Aue, Erzgebirge. *Center left* Native gold, × 4 approx. The crystals show platy to lumpy distortions. Verespatak, Romania. *Center* Arborescent native copper, Keeweenaw, Michigan. *Center right* Very much enlarged octahedron of diamond with etch figures clearly visible. Actual size of crystal (barely 0.2 carat) shown below. South Africa. *Below right* Coarsely foliated graphite. Ragedara, Ceylon.

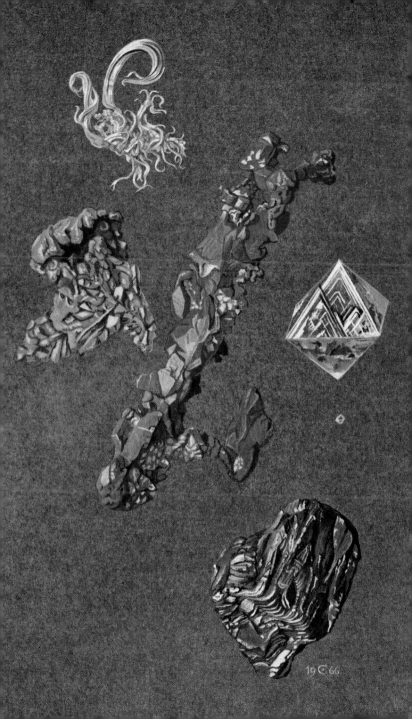

19 C 66

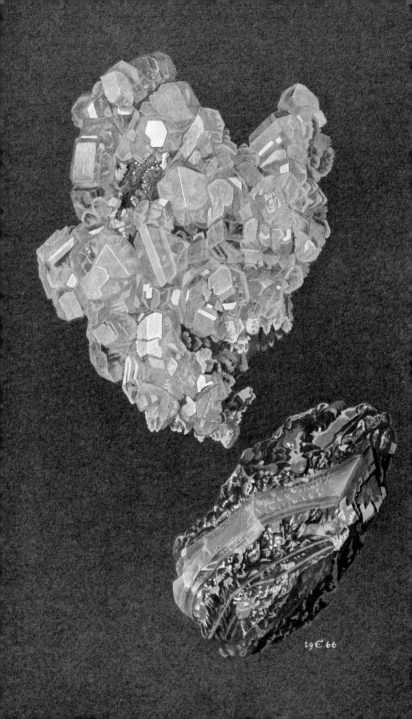

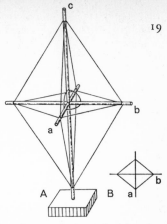

1 Orthorhombic axial cross (A). The three axes a, b, c are unequal in length and cut each other at right angles. The bipyramid outlined here has a rhomb-shaped cross-section (B).

CRYSTALS

The occurrence of well-developed crystals bounded by flat faces (for the definition of crystal see p. 11) often astonishes people, because straight edges and planes are usually regarded as characteristic of artificially manufactured objects, and naturally occurring substances are more usually associated with irregular curved lines and uneven surfaces. Irregular aggregates of minerals are of course by far the most common in nature. A collector should, however, know something about crystallography because not only are beautiful natural crystals the most desirable for the collector, but their form is a very characteristic feature of some minerals and provides valuable help in their identification.

All crystals may be classified into six systems. These systems are described by means of *imaginary axes* drawn through an ideal crystal from its center. These are called *crystallographic axes* or *axes of reference* and can be of equal or unequal length. Those of equal length are denoted by the letters a_1, a_2 and a_3. If they are of different lengths they are denoted by a, b and c. Depending on the system, the angles between the axes may be 90°, as in the cubic system, or some other value (see Table 2). Fig. 1 illustrates the axes of an orthorhombic crystal (and the outline of an orthorhombic bipyramid has been drawn in). The crystal systems, together with the corresponding basic geometric shapes, are explained and illustrated in Table 2. The name *cubic* is derived, of course, simply from the cube, and the names

PLATE 2   *Sulphur*
Above Sulphur crystals on calcite. *Below* A specimen of massive sulphur. Both from Agrigento, Sicily.

hexagonal (+ trigonal), *tetragonal* and *orthorhombic* from a cross-section of the crystal, i.e. six-sided, three-sided, four-sided and rhombic respectively. *Monoclinic* means that one of the axes in this system is inclined to the plane of the other two whilst *triclinic* refers to the fact that all three of its axes are inclined to one another.

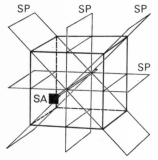

2 Cube showing four planes of symmetry (SP) cutting its front face. The black square SA indicates one of the fourfold axes.

Crystals are usually symmetrical objects. Fig. 2, for instance, shows how four *planes of symmetry* or '*mirror planes*' may be drawn through a cube, each dividing the cube into two equal parts which are mirror images of each other. There is also a fourfold *axis of symmetry* which is perpendicular to one of the cube faces and goes through its center. If the cube is rotated around this axis its initial position is reached again after each of four 90° turns. The same is true for the other faces of the cube, making three of these fourfold axes in all. The considerably lower symmetry of a monoclinic crystal is shown in Fig. 23 (p. 223).

The crystallographic symmetry of a body may often be lower than its apparent geometric symmetry. In a pyrite cube, for

TABLE 2 The crystal systems.

The diagrams show the six axial crosses, in each case illustrating a simple prismatic crystal form. For examples other than Figs. 4 and 5, see also the following figures and plates: *cubic* Pl. 5 tetrahedrite, Pl. 6 galena, Pl. 8 pyrite, Pl. 10 halite, Pl. 11 fluorite, Pl. 13 magnetite, Pl. 31 garnet, Pl. 43 leucite and analcite, Fig. 15 tetrahedrite, Fig. 16 pyrite; *hexagonal and trigonal* Pl. 12 corundum, Pls. 14–17 quartz, Pls. 22, 23 calcite, Pl. 30 apatite and pyromorphite, Pl. 37 beryl, Pl. 38 tourmaline, Fig. 17 quartz, Fig. 20 calcite; *tetragonal* Pl. 35 zircon, Pl. 36 idocrase; *rhombic* Pl. 27 barites, Pl. 34 topaz, Fig. 1; *monoclinic* Pl. 39 augite, Pl. 40 hornblende, Pl. 44 orthoclase, Pl. 45 adularia, Fig. 23 augite and hornblende, Fig. 24 feldspar; *triclinic* Pl. 44 microcline, Fig. 8 albite.

No of axes ------- Desig-nations	Interaxial angles	Axial length	System	Characteristic forms	Axial cross
3 ------- $1_1, a_2, a_3$	all three at right angles to each other	all three equal	*cubic*	cube, octahedron, rhombic dodecahedron, tetrahedron, pyritohedron	
4 ------- a_1, a_2, a_3 c	a_1, a_2, a_3 at 120 to each other, c at right angles to these	a_1, a_2, a_3 equal, c longer or shorter	*hexagonal* and *trigonal*	six-sided or three-sided prisms and pyramids, rhombohedra	
3 ------- a_1, a_2, c	all three at right angles to each other	a_1, a_2 equal, c longer or shorter	*tetragonal*	four-sided prisms and pyramids	
3 ------- a, b, c	all three at right angles to each other	all three different in length	*orthorhombic*	prisms and pyramids with rhomb-shaped cross-sections, pinacoids	
3 ------- a, b, c	two at right angles, the third inclined to these	all three different in length	*monoclinic*	prisms with inclined pinacoids	
3 ------- a, b, c	all three inclined to each other	all three of different lengths	*triclinic*	pinacoids only	

instance, because of the striations of the faces, the diagonal planes of symmetry are no longer present (Fig. 16a, p. 93). A lower crystallographic symmetry is often only shown by etch figures (pits formed in the faces by carefully treating the crystal with acids) or, sometimes, when the crystals are naturally dissolved after formation. Calcite and dolomite are thus both trigonal, but the true symmetry of dolomite is lower than that of calcite, as shown by different etch figures.

Some of the most important *crystal forms* are shown in Figs. 4 and 5. Their names usually end in *'hedron'* (surface); thus an *octahedron* is an eight-faced solid, a *tetrahedron* four-faced, a *dodecahedron* twelve-faced, and so on. It can be seen that some of the illustrations represent *closed forms,* in which the entire crystal is made up of one form only but other forms too are shown, especially prisms, which are *open*. To complete a crystal made up of prisms at least another pair of faces is needed. Crystals of pyromorphite (Pl. 30, p. 142) may be taken as an example of the hexagonal prism plus basal pinacoid. It is quite usual to speak of combinations of two or more crystal forms, but only forms belonging to the same system may be combined with each other. Combinations of four-sided prisms with six-sided pyramids or a rhombohedron are therefore impossible. Fig. 3 illustrates the formation of complicated combinations by two examples.

By looking through the plates and figures one soon sees that the majority of crystals consist not of single forms but of combinations of forms. See especially pyrite (Fig. 16, p. 93), quartz (Figs. 17 and 19 and Pls. 14–17), apatite (Pl. 30, p. 142), idocrase (Pl. 36, p. 164), augite and hornblende (Fig. 23, p. 223). If several forms with nearly equal angular inclinations occur, the crystals may often appear to be striated and rounded (see Figs. 3 and 16, Pls. 6 and 38).

Combinations of forms are frequently named after the predominant crystal form. It is therefore usual to speak of octahedral (galena, Fig. 3 and Pl. 6), tetrahedral (tetrahedrite group, Fig. 15 and Pl. 5), or prismatic (beryl, Pl. 37) *habits*. Some minerals occur in only one, or just a few characteristic habits, whilst the habits of others may vary a great deal: calcite, for instance, has over 80 known forms. This often makes it very difficult to recognize the crystal symmetry, so valuable for the

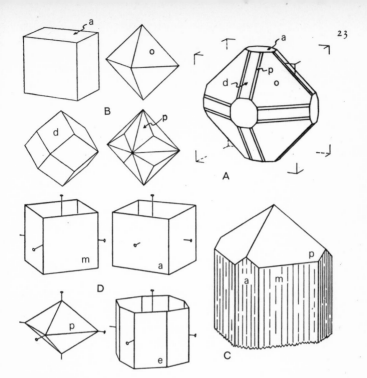

3 Crystal forms and combinations. (A) Cubic crystal made up of the closed forms shown in B. (B) Cube a, octahedron o, rhombic dodecahedron d, trisoctahedron p. The lettering in A and B indicates corresponding forms; the orientation of the cube in A is also shown by its corners. (C) Combination of two tetragonal prisms m and a (open forms) with a fairly flat bipyramid p, shown separately at D. The striations on the prism faces in C are a clue to the presence of a ditetragonal prism e, also shown at D (only the upper part of the whole crystal is shown).

identification of minerals. Further, and far more serious, difficulties arise from the fact that crystals are usually somewhat distorted; although the interfacial angles remain constant, the relative sizes of some of the faces with respect to others can vary considerably. In this way weird crystals may be formed, such as the distorted octahedron shown in Fig. 6 or the quartz specimens in Fig. 19, p. 133, where the basic forms may be almost beyond recognition.

Ideal crystals like those shown in Figs. 4 and 5 do not have *re-entrant angles*. This rule, however, is also frequently broken

24

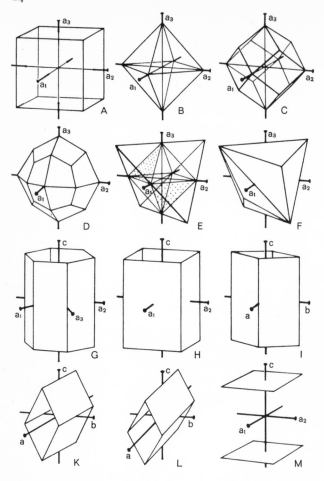

4, 5 Important crystal forms. (A) Cube. (B) Octahedron. (C) Rhombic dodecahedron. (D) Trapezohedron. (E) Tetrahedron derived from the octahedron by eliminating four of its faces (two of which are shown dotted). (F) Tristetrahedron; each tetrahedral face carries a flat three-faced pyramid. (G, H) Hexagonal and tetragonal prisms. (I, K) Vertical and horizontal orthorhombic prisms. (L) Inclined monoclinic prism. (M) Tetragonal basal pinacoid (orthorhombic, monoclinic and triclinic pinacoids may be horizontal, vertical or inclined).

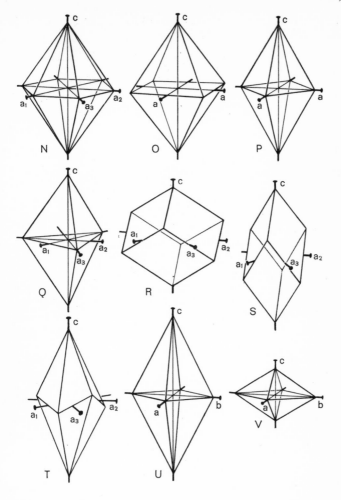

(N) Hexagonal bipyramid. (O, P) Second and first order tetragonal bipyramids. (Q) Trigonal bipyramid. (R, S) Normal and steep rhombohedra (note: 'rhombohedron' has no connection with the orthorhombic system). (T) Trigonal scalenohedron. (U, V) Steep and low orthorhombic bipyramids.

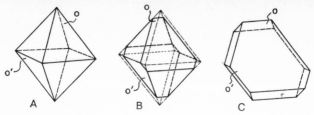

6 Example of a distorted crystal. The two parallel faces o and o' in A increase their size in B at the expense of the other faces, but all interfacial angles remain the same. (C) A very distorted octahedron in exactly the same orientation as in A and B. Gold sometimes crystallizes in such flattened octahedra.

by nature, but only apparently. If crystals with re-entrant angles are more closely examined, it is soon found that they are not single crystals, but two or more individual crystals intergrown and geometrically related to one another. *Twinned crystals* form one of the common groups of such intergrowths, where two or more identical individual crystals, turned with respect to each other, have grown side by side, or through each other, in a regular manner (contact twins, Figs. 7, 8, 21; interpenetrant twins, Figs. 16, 18; staurolite, Pl. 33). Depending on the number of individuals concerned they are referred to as *twins, trillings,* or *multiple* twins. Numerous thin lamellar crystals may alternate: Figs. 7 and 8 illustrate the formation of *spinel twins,* and lamellar twinning in albite, respectively. Twins are formed according to various 'twinning laws', named after crystals they exemplify, e.g. *Spinel Law, Albite Law* etc. (cf. also pyrite, p. 92; quartz, p. 127; gypsum, p. 177; feldspar, p. 242, etc.). A twinned crystal must always possess a different symmetry from the original (Fig. 8), frequently simulating a higher degree of symmetry, as is the case in aragonite or chrysoberyl (Pl. 12, (p. 64). In both cases orthorhombic crystals are twinned, forming apparently hexagonal trillings.

Although parallel twins may still have re-entrant angles, their crystals are no longer inclined towards each other, nor is there any additional symmetry generated (see calcite, Pl. 23, p. 119). The crystals may occasionally not be exactly parallel to each other but slightly inclined so that peculiar 'twisted' groups are formed (see 'twisted' quartz in Fig. 16, p. 93). In other instances it may be seen, after close inspection, that apparently single crystals in fact consist of a number of smaller crystals in not entirely parallel orientation; this may account

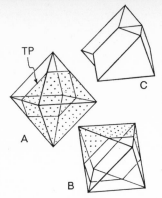

7 Spinel twin law. Rotating octahedron A with respect to octahedron B and joining the two halves together along the plane TP, makes the spinel twin shown at C. Dotted halves of the octahedra disappear. A new triangular shape replaces the original cubic symmetry, and the twin plane TP becomes a mirror plane.

for the so-called *mosaic* structure (see fluorite, Pl. 11, p. 63).

The external form of crystalline substances is directly related to their internal structure (see also p. 191). It is true that the smallest atomic building blocks differ depending on the substance in question, but they are always arranged in a regular array, termed *crystal lattices*. In the example provided by the simple cubic lattice of rock salt (Fig. 9) the lattice points are alternately occupied by sodium and chlorine ions. In quartz, each silicon atom is surrounded tetrahedrally by four oxygens and these SiO_4 tetrahedra are connected to each other in the form of a spiral.

A large number of such lattice types are known, including some very complicated ones. Some lattice types are characteristic of several different minerals and artificial compounds crystallizing in them. Halite (rock salt), sylvite, galena, and a host

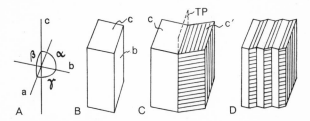

8 Lamellar twinning illustrated in albite. (A) Triclinic axial cross with interaxial angles of albite, i.e. α 94°, β 117°, γ 88°. (B) Idealized crystal of albite. (C) Two albite crystals, mirror images of each other, joined along b (the 'albite law'). The face c is striated in the twin face c′, and the mirror plane TP is the new symmetry plane. (D) This twinning process may be repeated many times, so that individual crystals become thin lamellae. These lamellae are often visible to the naked eye, as in labradorite (Pl. 46).

28

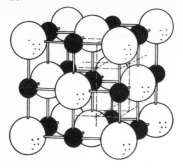
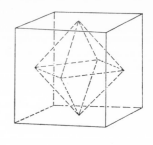

9　Atomic structure of crystals. *Left:* Model of the rock salt lattice. Na atoms are represented by black spheres, Cl atoms by white ones. The cleavage directions, along which the crystal will break under stress, are shown by broken lines. *Right:* Crystal forms corresponding to the positions of the white spheres (cube and octahedron).

of other compounds belong to the same lattice type. The internal arrangement of atoms within crystals may have a bearing on their external form or forms. In the rock salt lattice illustrated in Fig. 9, the squares, occupied by five white and four black spheres respectively, correspond to the cube faces. If only the six spheres occupying the centers of these cube faces are connected an octahedron is picked out. From a larger portion of the structure the rhombic dodecahedron and other geometric solids may be picked out in a similar manner. Other important characteristics such as *cleavage* (see p. 59) also directly arise from the structure of the lattice. Occupation of lattice planes by certain atoms may result in cleavages parallel to these planes, as indicated by the cubic cleavage of rock salt (Fig. 9). This connection becomes particularly obvious in the highly perfect flaky cleavage of the micas, based on a sheet-like arrangement of SiO_4 tetrahedra. The spiral structure of quartz, on the other hand, accounts for the absence of any good cleavage, since it is difficult to have any suitable planes cutting through a spiral. There are also other reasons for breakage: separation, or *parting,* often occurs along planes of weakness parallel to twin lamellae as in corundum (Pl. 12, p. 64), or parallel to planes of intergrowths with other minerals.

Although the crystal system, symmetry, and consequently the lattice type are characteristic of a given mineral, there are mineral compounds which can crystallize in two or more systems; these are said to be *polymorphs*. $CaCO_3$, for example, commonly occurs in the trigonal form as calcite but it may also occur as the orthorhombic mineral, aragonite. Strictly speaking these are two different mineral species. SiO_2 occurs in several different modifications, listed in Table 9 (p. 124), including also some *amorphous* forms where the basic building blocks are in a state of disorder. Carbon perhaps presents the most striking example of *polymorphism*, with the widely differing properties of graphite and diamond. Uusually one modification is common, while the others are rather exceptional. The common modification may also often have the same external form as the others. Leucite, for instance (Pl. 43, p. 203), is tetragonal at ordinary temperatures but cubic at higher temperatures, and crystallizes in well-developed trapezohedra (icositetrahedra). The external shape of the crystals is retained even after cooling whilst the internal structure, the lattice, resumes tetragonal symmetry. Such simulated forms are known as *paramorphs*.

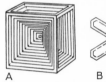

10   Hopper crystal formation. (A) Crystal of rock salt; growth along edges is faster than on faces. (B) Snow crystal, showing hexagonal symmetry.

A   B

Under certain conditions many minerals may imitate the crystal forms of entirely different minerals. These 'ghost' crystals are known as *pseudomorphs*. Pseudomorphs of limonite after pyrite are common, since pyrite readily decomposes into limonite (see p. 92). Malachite imitating azurite is a further example (p. 172). Dislocations in the lattice, growth faults and the like cause *mosaic* structures and *skeletal* formations, where growth along the edges and corners of a crystal is more rapid than on its faces (Fig. 10 or Pl. 16, p. 80, center).

CRYSTALLINE AGGREGATES AND VARIETIES

Although nearly every known mineral occurs in several different external forms, large well-formed crystals are relatively rare. *Crystalline aggregates* are made up of tiny, irregular crystal grains of widely differing shapes. Compositional variations, impurities, color changes, peculiar crystal forms, which may occur in a mineral, are often given separate names and are *varieties*. It was once common to give unnecessary names to minerals of distinct habits or from particular localities; corundum, for example, is also known as *sapphire* and *ruby*, beryl as *emerald* and *aquamarine*, and quartz has dozens of varietal names (Pl. 10, p. 58, but see also Pl. 9 again for the modifications of silica). However striking the external differences observed in one and the same mineral, ranging from well-developed distinct crystals to mere crystalline aggregates, they should not be stressed by giving them separate names. The name pyrite, for instance, applies to the coarse, unsightly ores as well as to the beautiful crystals shown in Pl. 8 (p. 52), and the massive pale yellowish-green veins in dark gangue rocks are just as much epidote as the famous dark-green, many-faced crystals from Knappenwand, Salzburg.

Massive apparently amorphous minerals are often termed compact. *Granular aggregates* are either coarsely or finely granular so long as the grains are still visible to the unaided eye (vein quartz, or quartz as sandstone); if they are only visible under a lens they are said to be of *massive* texture. The term *cryptocrystalline* means that under the microscope no individual crystals are discernible with any magnification (chalcedony, Pl. 18, p. 86). Aggregates of grains having distinct cleavage in various directions are 'sparlike' or *spathic* (calcite in marbles). In *platy, granular platy,* and *flaky* aggregates there is distinct cleavage parallel to a plane (micas, chlorites, bronzite). If crystals are noticeably elongated their aggregates are said to be *fibrous, radiating* (amphiboles), *acicular* ('needle-like'), *feathery* (stibnite), *bladed, lathlike* (kyanite, Pl. 33, p. 157), etc.

'*Radiating*' means that individual needles or fibres radiate from a common center (hematite or reniform limonite, Pls. 13, p. 73, and 21, p. 101). Easily crumbled minerals are described as *earthy, powdery,* or *mealy. Coatings* are thin or thick *encrustations* coating the surface of other minerals or rocks. *Hairs* or *wires, malleable sheets, lumps* and *nuggets* are self-explanatory (silver, copper). *Reniform* (kidney-shaped), *mamillary,* and *spherical* are used to describe hemispherical layered aggregates, consisting of radiating fibres (Pls. 13 and 21). *Columnar* formations are often said to be *stalactitic* (stalactites are pendent columnar concretions of calcite hanging from the roofs of limestone caves, and stalagmites are the corresponding formations growing upwards from cave-floors). 'Peastones' or *pisolites* and *oolites* are rounded bodies with concentric or radiating structures usually below 5 mm. in diameter, of calcite, aragonite, limonite, pyrolusite, or other minerals. Tabular minerals (gypsum, barite, hematite) may occasionally form rosettes ('desert roses'). *Dendrites* are moss-like limonite or pyrolusite found along minute flaws in rocks, and are mistaken for fossil plants.

2 The formation of minerals

Minerals occur everywhere in the Earth's crust. All rocks, including sand and even clay, consist exclusively of minerals. The individual grains of a rock sample are usually only a millimeter or so in size, frequently considerably smaller, sometimes unrecognizable even under strong magnification. All this does not excite the collector and he needs to find out where the relatively rare, larger single minerals are to be found. Natural mineral aggregates, of course, do not consist of just any kinds of minerals, but under particular conditions certain combinations may be found over and over again as regular associations. These are characterized by the common origin of the minerals they contain and are known as *parageneses*, i.e. 'formed together'. Mineral parageneses originate in several ways. Crystallization from natural melts is perhaps the most important, as can be observed directly in lava flows after volcanic eruptions. Minerals also crystallize from hot or cold solutions. During the weathering of minerals new ones are formed, and certain living organisms may also contribute towards their formation. Feldspar, quartz and mica, for instance, is a very common association, forming the rock granite; in mineral deposits galena is almost invariably accompanied by sphalerite. Certain other minerals may exclude one another. Thus large amounts of feldspar are hardly ever found together with halite. Some minerals can be formed under very variable conditions, and occur in several different associations; these are called *gangue* minerals. Only by learning about the many different associations of minerals is it possible to get to know something about their occurrence in general and where well-developed crystals are likely to be found.

A definition of a rock has already been given (p. 10); a rock is a natural formation consisting of one or more minerals with general distribution. Rocks must be geologically significant masses. The characteristic rock-forming minerals are the chief constituents of rocks, but they also contain small quantities of various accessory minerals which hardly affect their over-all compositions. Mineral deposits are distinct from rocks. These are concentrates and may consist of rock-forming and/or accessory minerals, but more often contain minerals not commonly found in the usual rock types. Although the general definition of rocks could normally also cover deposits, there are many good reasons for the distinction. Deposits are often formed under very special conditions, mainly due to the distribution of the chemical elements. Table 3 lists the fifteen most abundant elements present in the upper ten miles or so of the crust, which account for over 99 per cent of the total weight. This leaves less than 1 per cent for all the other 75 or so elements. The rock-forming minerals consist almost entirely of these fifteen elements alone (Table 4).

It is obvious that some important elements and useful metals do not occur in any of the assemblages. Cu accounts for only 0.008 per cent, Zn 0.007 per cent, Pb 0.002 per cent, U 0.0004 per cent and Au 0.0002 per cent of the Earth's crust. If the rare elements were distributed equally they would be completely inaccessible. In fact minerals containing most of these rare elements do occur in nature in varying degrees of concentration, or at least in sufficiently enriched zones to enable their detection and commercial working.

Before dealing with the actual formation of minerals and the chemical composition of rocks and ore deposits on which their classification is based, a few of their external characteristics

TABLE 3 The 15 most abundant elements in the upper 10 miles of the Earth's crust given as weight percentages (rounded off)

oxygen	O	47.3	potassium	K	2.8	phosphorus	P	0.1
silicon	Si	30.5	sodium	Na	2.5	fluorine	F	0.1
aluminum	Al	7.8	magnesium	Mg	1.4	hydrogen	H	0.1
iron	Fe	3.5	titanium	Ti	0.5	sulphur	S	0.05
calcium	Ca	2.9	manganese	Mn	0.1	chlorine	Cl	0.05

TABLE 4 Common rock-forming minerals (including some accessory minerals as well)

Class 1	graphite
Class 2	pyrite, pyrrhotine
Class 3	halite
Class 4	quartz, magnetite, hematite
Class 5	calcite, magnesite, dolomite
Class 6	anhydrite, gypsum
Class 7	apatite
Class 8	olivine, andalusite, kyanite, staurolite, garnets, epidote, zircon, tourmaline, cordierite, sillimanite, augite, hornblende, talc, muscovite, biotite, chlorite, serpentine, clay minerals, orthoclase, plagioclase, nepheline, leucite.

must be mentioned. Examination of hand specimens of different rocks or ore samples presents a pretty complex picture regarding the shape, size, and arrangement of mineral grains. Individual grains may all be of roughly equal size or we may find larger grains of one kind in a finer ground-mass of other kinds. Minerals may be present as irregular grains, in tabular, platy, flaky, or fibrous forms, occasionally even in crystals bounded by smooth faces. The components may form a uniform mixture, or the same minerals may appear in parallel, regularly alternating, intergrowths. Platy or fibrous minerals may be arranged in special orientations. Ores may equally consist of well-ordered bands of minerals and completely irregular forms (lenticles, schlieren, lumps, pockets). This outward appearance is usually referred to as *texture*.

Rocks and ore-bodies may exhibit a variety of shapes depending on their origins: varying sized bosses (stocks), veins, pipes, plugs, uniform or alternating layers, smaller lenticles or larger lenses or completely irregular masses. However, subsequent movements within the crust, the so-called *tectonic* processes frequently destroy or damage these original formations; they are fractured and displaced along fracture planes, relatively raised or lowered, even pushed on top of one another. At sufficiently high pressures these seemingly rigid mineral bodies become plastic and can be bent, or *folded;* some stratified rocks show this most clearly. The resulting new cracks and cavities often lead to the formation of new minerals. These tectonic processes may be responsible for uplifting vast portions of the crust to form mountain ranges, exposing thereby, for the first time perhaps, deeper parts of the crust.

34

ROCKS

According to their origin rocks can readily be classified into three main groups: *igneous; metamorphic; sedimentary*. They may then be subdivided further first by their origin and then by their mineral content.

Igneous rocks (known also as *magmatic* rocks) are classified partly by their mode of solidification from the molten state, and partly by their composition, i.e. mineral content (Fig. 11 and Pl. 5, p. 41). *Acid* rocks consisting largely of silicon-rich and iron-poor minerals (quartz, potash and soda feldspars) can thus be distinguished from *basic* types characterized by minerals with relatively low silicon and considerably higher iron content (calcic feldspars, hornblende, augite, olivine, iron ores). The so-called *intermediate* rocks lie between these two. A more detailed classification is based on the mineral content of individual rocks.

The natural process taking place, illustrated in Fig. 11, may be visualized as follows: part of a deep-seated original liquid *magma* triggered off by geological disturbances penetrates into the upper regions of the Earth's crust and there begins to cool slowly. The 'basic' or heavy minerals are the first to crystallize and sink to the bottom *(differentiation)* where they may often be redissolved. If their iron content happens to be particularly high, droplets of iron oxides or iron sulphides are formed and in turn sink, collecting again at the bottom of the main mass of magma (liquid magmatic *segregation*). These processes may radically affect the composition of the magma in certain zones, so that the originally uniform magma is split into two parts, an upper one of granitic composition and a deeper-lying part of gabbroic composition.

These are separated by a layer of intermediate composition (compare the mineral contents of granite, gabbro, and diorite in Table 5). When completely solidified such a magma becomes a *plutonic* rock. Some portions of the melt may have intruded into the country rock along crevices and cracks and solidified

PLATE 3   *Sphalerite*
Above Black, iron-rich crystals of sphalerite with brownish dolomite and calcite. Trepča, Yugoslavia. *Below* Fine-grained fibrous sphalerite from Diepenlinchen, near Aachen, Germany.

37

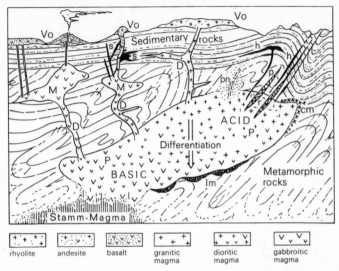

rhyolite   andesite   basalt   granitic magma   dioritic magma   gabbroitic magma

11   Schematic representation of processes taking place within the Earth's crust, to explain the formation of igneous rocks and some ore deposits. P, plutonic rocks; D, intrusive rocks; Vo, volcanic rocks; M, near-surface magma reservoirs; lm, liquid magmatic ore bodies; r, residual crystallization; p, pegmatites; pn, pneumatolytic products; h, hydrothermal veins and replacements; cm, contact-metamorphic alterations in the country rock.

there, forming so-called *intrusive* rocks. Their composition corresponds to the magma they are derived from. There are also minor intrusions, which do not represent the normal, entire mineral content. *Aplite* and *pegmatites* are such intrusions, consisting largely of lightly colored minerals (more fully discussed under Ores, p. 44), while the lamprophyres predominantly consist of dark minerals. Lamprophyres have been given a host of other names and this is the reason for omitting them from the table.

Should the molten mass penetrate to the surface, volcanic phenomena will occur. The resulting extrusive rocks may either form extensive sheets, cones or complete volcanoes of

TABLE 5 The principal igneous rocks

K, Na or Ca after feldspar stands for predominant potash, soda or calcium feldspar respectively. Geographically old rock types are marked *

mineral content	plutonic rocks	volcanic rocks	dyke rocks	split dyke rocks
quartz feldspars (K, Na) mica	granite	liparite rhyolite *quartz-porphyry	granite-porphyry	aplite-pegmatite
quartz feldspars (Ca, Na) hornblende	quartz-diorite	dacite	quartz-porphyry	aplite
feldspars (Na, Ca) hornblende	diorite	andesite *porphyry	diorite-porphyry	diorite-aplite
feldspar (K) hornblende	syenite	trachyte *porphyry	syenite-porphyry	syenite-aplite
feldspars (K, Na) nepheline hornblende augite	nepheline-syenite	phonolite	nepheline-porphyry	nepheline-aplite nepheline-pegmatite
with or without olivine feldspar (Ca) hornblende, augite iron ores	gabbro norite	basalt *dolerite (diabase) *melaphyre	gabbro-porphyry dolerite (diabase)	gabbro-pegmatite
olivine augite hornblende iron ores	peridotite dunite	picrite limburgite		

lavas and ashes. The word *lava* does not refer to any specific rock types. Any magmatic flows or streams solidified on the surface of the Earth's crust in contact with air, irrespective of composition and type, are *lavas*. Near-surface volcanic rocks are often accompanied by *explosion-breccias* which also contain some country rock (see Fig. 12). If, during a volcanic eruption, some molten magma together with some of the country rock is hurled through the air, *tuffs,* ranging from coarse to very fine material, are formed.

The sequence of events occurring within the Earth's crust described here is oversimplified and only serves to illustrate the principles behind the processes. A magma may actually re-melt the country rock and thereby considerably alter its composition, and the sequence of events in this case is very complex. In addition a number of plutonic rocks, especially granites, may have had a different origin altogether. If, due to ever-increasing superimposition, already existing rocks begin to sink once more, they are altered (see p. 43), even re-melted if they sink down sufficiently far, and eventually fresh magma is formed. It must not be forgotten that only surface volcanic activity may be observed directly, whereas plutonic and minor intrusive rocks are exposed only after weathering.

Plutonic rocks are usually evenly and coarsely crystalline, whilst basic intrusive and volcanic rocks are finely crystalline to massive. Intrusions and lavas of acid to intermediate composition have so-called *porphyritic* textures: larger crystals are suspended in a very fine ground mass, in which individual crystal grains are no longer recognizable. The ground mass may sometimes consist entirely of volcanic glass, especially if the melt, in which already fully developed crystals are suspended, cooled down too rapidly for crystallization to take place. Lavas which consist entirely of glass are known as *obsidian.*

Sedimentary rocks may be considered as erosion products of other rocks. Erosion may have been partly mechanical (by the action of wind, water, frost etc.), the rocks being first loosened and then broken up, and partly chemical (by water containing carbon dioxide dissolving or decomposing the rocks). Sedimentary rocks are classified according to their formation and composition, in some cases also according to grain-size. They fall into two main groups: if the eroded rock material stays at its original site or is rearranged only slightly we talk of *residual* or *weathered* rocks. If, on the other hand, as usually happens, the eroded material is carried further afield either in solutions or mechanically, by water, wind, or glaciers, and redeposited, real *sedimentary rocks* are formed. Deposition may take place on land, in rivers, lakes, or in the sea. The majority of sediments are *marine,* i.e. deposited in the sea.

Residual formations, sediments and even loose grains such as sand or gravel are geologically still regarded as rocks. Such loose grains are eventually compacted either by pressure or by deposits from solutions cementing together individual grains.

TABLE 6 The sedimentary rocks

1   RESIDUAL ROCKS

grit, gravels (in part), weathered clay, laterite, bauxite

2   SEDIMENTARY ROCKS

Chemical and organic deposits
I   pure chemical sediments: some marine limestones and dolomites, calcareous tufa (in part); halite (rock salt) with gypsum and anhydrite; manganese sinter, iron oolites
II chemical-organic deposits: main mass of marine limestones, reef formations, chalk, shelly marls, calcareous tufa (in part); radiolarite, hornstone, siliceous sinters, diatomaceous earths
III wholly organic sediments: coal, bitumen, oil

Mechanic or clastic sediments
I   psephites (grains larger than 2 mm.): grits and breccias (angular grains), gravels, conglomerates (rounded grains), moraines and tillites (mixture of very coarse and very fine-grained matter)
II psammites (grain size 0.02–2.0 mm.) sands and sandstones: arkose (quartz, feldspar), greywacke (quartz, feldspar, clays), quartz sandstone (quartz)
III pelites (grain size below 0.02 mm.) clay, argillite, loess

This process, known as *diagenesis,* often also involves the formation of new minerals (e.g. crystalline aggregates).

Sediments are extremely variable (Table 6) and numerous. Pure chemical sediments are formed by 'chemical' processes, like the evaporation of water containing dissolved substances; salt deposits are an example. Evaporation by itself is a physical process, of course, but the resulting sediments have always been regarded as chemical; *organogenic* rocks are formed if the dissolved substances are removed from the water by living creatures. The formation of shells and skeletons (mussels, snails, sponge spicules etc.) from calcite or silica belongs here. Some organisms even build their own external skeletons (corals, calcic algae). Extensive limestone deposits can be almost entirely made up from the shells of microscopic creatures, and massive reefs owe their size to the activity of reef-forming plants and animals. Pure organic deposits are formed when the actual organic matter of plants (forming coal) or

PLATE 5   *Tetrahedrite*

Above Tetrahedrite crystals on barite, showing a rhombic dodecahedron, with modified edges, on the left. Brixlegg, Austrian Tirol. *Below* Tetrahedron of tetrahedrite coated with chalcopyrite, with foliated siderite and calcite crystals. Clausthal, Harz Mts., Germany.

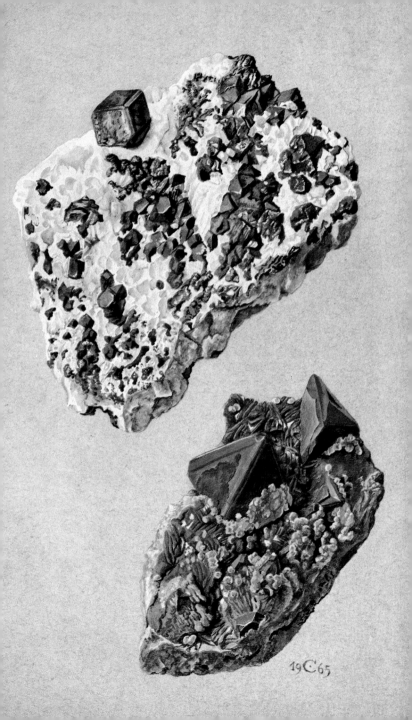

animals (bitumen, oil) has become rock. Mechanical or clastic sediments are formed when solid particles, like rock fragments, sand, or clay minerals are transported and then deposited.

Some rocks cannot be definitely assigned to any one of these groups. If chemical deposition of chalk occurs simultaneously with clastic deposition of clay, *marls* are formed. Some dolomites must have formed from limestones by the addition of magnesium during diagenesis. When material from a volcanic eruption gets deposited on the sea-bed, *tuffs* are formed. These should really be classed as igneous rocks because of their origin.

The group of *metamorphic* rocks includes all rocks considerably altered by high temperatures or pressures some time after their formation. Recrystallization into new minerals, brought about by unilateral directed pressures is a chief characteristic of metamorphism. Flaky crystals particularly, but fibrous and others too, arrange themselves in layers, giving the rocks a *slaty texture*. For this reason the metamorphic rocks are called *crystalline slates*. (Some thin-layered sedimentary rocks, however, can also be called slates.) Diagenesis is not a metamorphic process even though temperature and pressure play a part and new minerals are produced.

Classification is based first of all on the degree of metamorphism in relation to the depth of the rocks in the Earth's crust:

1 Contact metamorphism:
   high temperature, high, all-round pressure
2 Regional metamorphism:
   Epizone: moderate temperature, directed high pressure, moderate depth.
   Mesozone: medium temperature, higher directed pressure, medium depth.
   Katazone: higher temperature, decreasing directed and increasing all-round pressure, great depth.

Contact metamorphism is the alteration, caused by heat, in a rock which is in direct contact with an intruding magma. The

PLATE 6   *Galena*
Large crystals of galena with brown rhombohedrons of siderite and rock crystals. Typical association from Neudorf, Harz Mts., Germany. *Below* A single crystal taken from this specimen; for habit, see Fig. 5, p. 25.

changes are localized; the so-called metamorphic aureole is from a few cms. to 1–2 kms. thick (see Fig. 11). Garnet-rock, hornfels, calc-silicate rock, slate or marble may be formed with beautiful large crystals of garnet, idocrase, andalusite, corundum, spinel etc.

Regional metamorphism, caused by positional changes within the crust, covers a wide area, or region. The slaty texture is very marked in the epi- and mesozone, while the rocks in the katazone may have textures reminiscent of plutonic rocks. If the extent of metamorphism exceeds the requirements of the katazone, remelting occurs and the rock becomes a magma again (p. 34). Depending on the original rock, there are *orthorocks* formed by the alteration of magmatic rocks, and *pararocks* which were originally sediments. Given the right composition the same metamorphic rocks may be formed from sedimentary rocks as from magmatic rocks.

Table 7 lists just a few examples. There are a vast number of metamorphic rocks, depending in turn on the number of original rocks and on differences in degree of metamorphism. Metamorphic rocks are often very rich in particular groups of minerals and are named accordingly: garnet-mica-schist, feldspar augen gneiss, sillimanite or cordierite gneiss, garnet amphibolite etc. Further kinds are: talc schists, frequently with crystals of magnesite (epizone); chlorite schists with crystals of magnesite and magnetite (epizone); epidote schists (epi- to mesozone); sericite, quartz and calcareous slates (epizone); granulite (katazone) etc.

ORE DEPOSITS

Ore deposits may be divided into three main groups, like the rocks with which they are associated. Some exceptional deposits do not readily fall into any of these. Differences of opinion may also exist as to their formation.

The ores of the magmatic zone are usually tied to the succession found in plutonic rocks. Only magnetite, titaniferous magnetite, chromite, and pyrrhotite are formed from liquid magmas, sometimes in enormous masses like the pentlandite-pyrrhotite deposit of Sudbury, Canada. The Sudbury ore is concentrated at the bottom of a norite trough 60 km. long. In such cases as the Swedish magnetite deposits at Kiruna, the ores are thought

Table 7 Some regional metamorphic rocks

The minerals in italics represent the constituents of the main rock, e.g. granite is composed of *feldspar, quartz* and *mica*.

Original rock	Epizone	Mesozone	Katazone
limestone *calcite*	marble *calcite*	marble *calcite*	marble *calcite*
quartz-sandstone *quartz*	quartzite *quartz*	quartzite *quartz*	quartzite *quartz*
coal	graphite *graphite*	graphite *graphite*	graphite *graphite*
bauxite *aluminum hydroxide*			emery *corundum magnetite*, etc.
arkose *quartz feldspar*	sericite-slate *sericite quartz*	mica schist *mica quartz little feldspar*	gneiss *feldspar quartz mica*
granite *feldspar quartz mica*	sericite-gneiss *sericite, quartz a little feldspar*		
marl *clay minerals calcite*	green glauconite schist *chlorite sericite*	amphibolite *hornblende feldspar*	eclogite *augite garnet*
basalt, dolerite *augite feldspar hornblende*	hornblende *a little feldspar*		
peridotite *augite olivine*	serpentinite *serpentine*		

to have become separated from the original magma after segregation and then to have been intruded as special 'ore-magma'.

Towards the end of the cooling of a granitic magma there may be a concentration of volatile substances (gases, water) in its upper regions, forming residual melts (Fig. 11) which are sometimes rich in rarer elements. These solidify as pegmatites in veins and small stocks by *residual crystallization,* often with

46

many well-crystallized rare minerals. All the largest known mineral crystals (such as feldspars of several hundred tons) are formed in this way. Pegmatites are important economic sources of many minerals, such as the feldspars, muscovite, beryl and gemstones.

If the residual melts contain a large proportion of gaseous components, *pneumatolytic* mineral assemblages *(pneuma,* gas; *lysis,* dissolution) develop with cassiterite, wolframite, and fluorine-bearing minerals such as topaz, tourmaline, and many others (see p. 143). Mineral formation from gases near the surface, as from volcanic emanations, is known as *sublimation* (direct transition from the gaseous to the solid state). Sulphur deposits are formed in this way, and occasionally workable veins of cassiterite with hematite (Mexico); many other minerals too form as sublimates but mostly in small amounts.

If the temperature drops to about 400°C during recrystallization, hot aqueous *hydrothermal* solutions predominate and the pegmatitic and pneumatolytic mineral assemblages gradually change into hydrothermal ones. Hydrothermal deposits near the surface are sometimes termed *subvolcanic*. Pneumato-

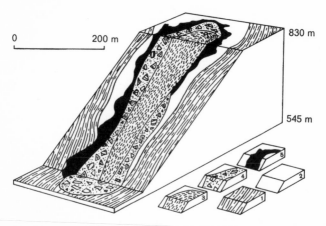

0   200 m

830 m

545 m

12 Diagrammatic block representation of the Stari Trg lead-zinc ore deposit near Trepča, Yugoslavia, an example of a hydrothermal subvolcanic replacement deposit associated with a volcanic pipe. The ore replaces the surrounding limestone. 1, Mica schist. 2, Limestone. 3, Andesite pipe. 4, Mantle of volcanic breccia surrounding 3. 5, Ores. (After Friedensburg, *Metallischen Rohstoffe*, 1950.)

lytic and hydrothermal processes produce ore veins (Fig. 13), then *replacements* or *metasomatic* bodies, where rocks, mainly limestones, are replaced by other minerals, and also *impregnations,* where any rock becomes finely interspersed with minerals which are not its normal components. The famous silver-rich ores in the Harz mountains, Germany, or the cobalt-nickel-silver-uranium veins in the Erzgebirge, Germany and Czechoslovakia, are hydrothermal. Trepča, Yugoslavia, known for its beautifully crystallized minerals, is an example of hydrothermal lead-zinc replacements (Fig. 12).

Pneumatolytic replacements often overlap with contact metamorphism (p. 43) when the latter is coupled with a supply of hot material from the melt; the combined effect is known as *contact metasomatism.* The ores of Broken Hill, Australia, are of this type and probably represent the largest lead-zinc concentration in the whole world. The so-called *disseminated ores* (i.e. finely divided ores) are very important hydrothermal impregnations, and occur where acid volcanic rocks are impregnated by molybdenite (Climax, Colorado), or cassiterite. When the country rock, usually granite, is pneumatolytically impregnated

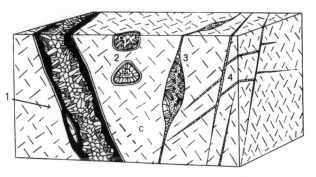

13   Block diagram illustrating an ore vein, druses, and alpine fissures. 1, Ore vein; this hydrothermal vein is symmetrically lined by ores (black and shaded) and filled with gangue minerals (white), such as quartz and calcite. Along the contact with the country rock there is an alteration zone. Veins may vary from a few cm. to several metres in thickness. 2, Druses, irregular to rounded cavities with crystals lining the walls. 3, Alpine fissures: thin cracks in the rocks widen to large cavities with room for free-standing crystals to grow in. 4, Crevices, a few mm. to a few cm. wide, often occurring in swarms: usually filled with quartz or calcite. c, Country rock.

by cassiterite and other minerals, becoming considerably decomposed in the process, the resulting deposit is a cassiterite *greisen*. Other examples of hydrothermal products are agate amygdales with amethyst, calcite, and other minerals in basalts; minerals deposited from hot springs, such as at Steamboat Springs, Nevada, where cinnabar, chalcedony, and even gold are continually formed at about $80°C$; and aragonite from the hot springs at Karlovy Vary, Czechoslovakia.

Lastly the famous mineral associations of *alpine fissures* must be mentioned. Their formation differs from other hydrothermal processes in that they are not residual solutions from melts, but the hot solutions derive their material from the country rock, depositing large crystals in cavities and druses (Fig. 13). The chief mineral is rock-crystal.

The ore deposits of the sedimentary zone are divided, as are rocks, into weathering products and true sedimentary ores. Weathering can result, depending on the original rock and the climate, in deposits of limonite (some laterites, see p. 121, and pea ore), aluminum (bauxite, see p. 152), manganese (p. 144), nickel (garnierite), kaolin, and other minerals. Many low-grade iron deposits may be considerably enriched by weathering, as itabirites and taconites. The weathering of sulphide ores (Fig. 14) is extremely interesting mineralogically. In the uppermost part of such deposits, provided that surface waters containing carbon dioxide and oxygen have access, most of the valuable metals (such as silver and copper) are dissolved by decomposition of their minerals and transported downwards. Iron is left behind, with all iron minerals altering to limonite; 'iron hat' or 'gossan' are miners' names for this *oxidation zone*, which can be found in every iron deposit, and in ancient times used to be worked preferentially. Lead is often not attacked, and copper and zinc in the presence of lime are often preserved as malachite, azurite, or smithsonite; the occurrence of these minerals with cerussite is famous in the oxidation zone at Tsumeb, S W. Africa. The oxidation zone, however, is impoverished in contrast with the *cementation*, or *secondary enrichment zone*, which begins where descending rich solutions first come into contact with underground water. Replacement of the weathered *primary* sulphide minerals takes place, forming *secondary* enriched deposits of chalcocite and sometimes of argentite. The

14 Section through the cementation and oxidation zone of a
sulphide deposit. c, Country rock.
W, Water table. 1, Primary ore.
2, Cementation zone. 3, Oxidation zone (gossan). Secondary
minerals are shown in black.

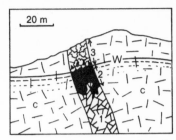

rich ore zones of the famous old mining localities are largely exhausted today, such as the silver of Mexico and South America, and the copper of Rio Tinto, Spain. Frequently, deposits of finely divided sulphides such as the disseminated copper ores are only workable where they have been concentrated by secondary enrichment.

Gravels are partly residual formations and partly clastic sediments. When rock is broken down by weathering, some minerals, mostly the heavy ones, remain unaffected. These are concentrated, and enrichment occurs when the other components are carried away (by rain or rivers). Gravel minerals may remain *in situ* or may be redeposited by rivers; tidal action forms gravel beaches. Gravels may be loose or recemented, and they are formed in every geological system. Gold, platinum, diamond, ruby, sapphire and other gemstones, rutile, ilmenite, magnetite, and cassiterite (three-quarters of the world's tin is derived from gravels) are important gravel minerals. The Witwatersrand gold deposits, South Africa, are a famous example of a cemented gravel (or conglomerate).

The chemical-sedimentary ores fall into two groups: most of them derive their constituents from weathering processes, but direct contribution of volcanic material plays an important part in some.

Purely sedimentary metal enrichments are the marine deposits of oolitic limonite, hematite and manganese (oolites are concentrically layered spherules from 0.5 mm. to a few mm. in diameter), clay ironstones, bedded manganese and copper. The metals are carried away from the land mass in solution and deposited in the sea. There are also freshwater deposits of iron, known as bog iron ore and lake iron ore. Salt deposits are characteristic examples of chemical sedimentation. Freshwater salt deposits (salt lakes and beds) are insignificant compared with marine deposits. The formation of marine salt deposits, often several hundred meters thick, with their intricate succession of halite and other minerals, is very complex. A simplified approach is to think of a marine lagoon, cut off from the open sea, but with a steady supply of sea water, where continuous evaporation occurs under arid climatic conditions. Calcite is deposited first, followed by gypsum, anhydrite, halite, and potash salts. Because of changing conditions this sequence of deposition is repeated, and bands of clay prevent the re-solution of salts. The sequence often finishes with halite or gypsum. Salt deposits are very widespread and often of enormous extent. Large amounts of material can be introduced into sea water during submarine volcanic explosions or gaseous effusions, and deposited later. Some of the sedimentary hematite and pyrite deposits are formed in this manner: an example of the latter is the deposit of massive pyrite, with chalcopyrite, sphalerite, galena, barite etc., at Rammelsberg, near Goslar in the Harz.

Finally there are minerals which crystallize from solutions present in rocks, where they occupy small cracks and fissures; an example is the formation of calcareous sinter and stalactites in cavities in limestone. *Concretions* in sediments may be formed before, during, or after compaction into a rock. They are irregular in shape, or they may tend to spherical; examples are the clay ironstones (p. 155) formed while marine clay is being deposited, or flint nodules, or calcareous nodules in loess. Circulating solutions may even lead to true replacements, as in some siderite deposits. These waters may also dissolve material from ore deposits and redeposit it after carrying it for a shorter or longer distance. Minerals so formed usually remain near the surface, or at least cannot be traced to great depths.

The *ore deposits of the metamorphic zone* are generally less important. Minerals present in ores may be altered in the same

PLATE 7   *Stibnite and cinnabar*
Left Stibnite. Both crystals are longitudinally striated, the front one being slightly twisted. Terminations missing. Shikoku, Japan. *Right* Massive cinnabar ore. Almaden, Spain.

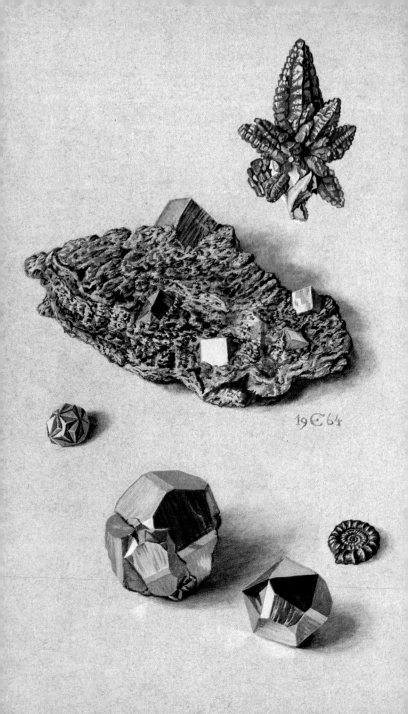

19 C 64

way as those present in rocks without any significant enrichment taking place. The extensive taconite and itabirite iron deposits, for example, were formed by metamorphism from sedimentary deposits. Some occurrences of manganese ores seem to have similar histories.

The geological history of the highly metamorphosed *skarn* ores is very complex, and they are restricted to old rocks. These deposits were partly sedimentary and partly hydro-thermal or pneumatolytic in origin, and have undergone several metamorphic changes accompanied by extensive re-crystallization. Material supplied by subsequently intruded igneous rocks plays an important role. Magnetite, or pyrite together with several other sulphides, are the principal ore constituents which, depending on the country rock, are associated with various silicates such as garnet, epidote, horn-blende, augite, serpentine, and olivine. The skarn ores of Falun and Boliden in Sweden have great economic importance.

3 Collection and identification of minerals

TYPES OF COLLECTIONS

When starting a collection it is usual to take anything within reach. If one's interest is maintained it will soon be desirable to introduce some order and to have a certain aim, but amassing a collection does not depend solely on personal inclination. Among the external factors, the amount of space available to an amateur for his hobby is of primary importance. If space is very restricted it may be possible to concentrate on certain localities, or even on individual crystals – but this is rather expensive. *Micromounts,* tiny fragments or crystals which are usually

PLATE 8 *Pyrite and marcasite*
Above Arborescent arrowhead intergrowth of marcasite. *Center right* Pyrite cubes in gneiss, the larger cube showing striations (*cp*. Fig. 16, p. 93). Habachtal, Salzburg, Austria. *Center left* Iron cross twin of pyrite. *Below left* Irregularly intergrown crystals of pyrite, combining cube, pyritohedron and octahedron. Elba, Italy. *Below center* Combined octahedral and pyritohedral faces. Elba, Italy. *Below right* Pyritized ammonite. Nürtingen, Württemberg, Germany.

mounted on small pedestals in special boxes, are becoming increasingly popular; although it is easy to store such a collection, it requires a microscope to look at it. The area in which one lives is also a controlling factor, because some regions are abundantly supplied with interesting minerals while others have very few or none. If good occurrences are near at hand, they may provide the stimulus for starting a local collection. One can also concentrate on one or a few particular mines but only if these are being worked at the time and the management is friendly. A *systematic collection* is good to have, but it will never be complete and success is more likely if one concentrates on the most important minerals. One should endeavor to obtain these in various habits, to represent several localities, and to include some massive samples as well as crystallized ones. This has its own inducements, because it will not be necessary to spend a lot of money and energy in chasing after showy specimens, and each ore sample will be a specimen for the collection. Even if only 50–100 species can be collected in this way, there are inexhaustible possibilities since even quartz or calcite alone occur in dozens of forms. Such a systematic collection can be extended at leisure to include the rock-forming minerals in hand specimens of rocks, with rare species added later.

STORING AND LABELING

Office cabinets with sliding drawers may be used for storage. Cupboards with drawers below and glass showcases above are of course very much better, especially for showing off and housing large specimens, but these may have to be custom-built. Sufficient drawer-space is essential if a collection is to be kept in reasonable order. It is equally important to have the specimens properly labelled and numbered; unfortunately, this is seldom done. Consecutive numbering is by far the best, and this may be done either by sticking on tiny labels, or by painting on a small colored spot and then writing the number on in india ink and covering with colorless lacquer. A catalog must be kept where everything about a specimen is entered under its number. The most important part of this information is the name, any associated minerals on the specimen, the chemical composition, special notes, and the exact locality where the specimen was found, giving as much detail about it as you can (wherever possible with map and literature references). The date of collecting, the name of the collector, and details of purchase or exchange should also be recorded. The more carefully a collection is assembled the greater its value. Specimens are often housed in suitable card trays together with their labels, but even so the specimens ought always to be numbered as well. Small fragments are best kept in glass or plastic tubes. Water-soluble substances should be specially protected, and are best stored in sealed glass containers after careful drying.

COLLECTING AND PREPARING

Not many things are needed for actual collecting: a short heavy sledge, a selection of chisels, a mineralogist's pick, a folding spade, a rucksack or bag for carrying, newspapers, soft tissue for wrapping, a few small boxes for any delicate specimens, paper or labels for rough notes, and, never to be forgotten, a note-book. The most expensive piece of equipment is a hand lens with 10 × magnification.

Quarries and mines are the best places to visit, but it is usually advisable to obtain permission. It will often be possible to acquire specimens from quarrymen and miners. All but the most delicate specimens should, as far as possible, be trimmed to their final size in the field (it will be helpful to get some practice in trimming beforehand on worthless material). Cleaning is best done at home. Rough and massive minerals and crystals may easily be cleaned with the help of detergent, water, and brush, but for the more delicate specimens, particularly the softer ones (like crystals of gypsum), it is best to use a soft shaving or artist's brush. Very delicate specimens, very fine needles for instance, are just rinsed in water. Chemically active cleansing agents are not advisable. Sulphide minerals should never come into contact with water; it is best to rinse them in alcohol.

IDENTIFICATION FROM EXTERNAL CHARACTERISTICS

The identification of minerals is not always easy. Unless he confines himself to the most common minerals such as quartz, feldspar, fluorite, calcite, pyrite, etc., a beginner will find the help of an experienced mineral collector invaluable.

How to identify minerals with the help of the tables in Part 3 is described below, and their external characteristics discussed. One begins by determining the *luster, streak,* and *hardness.*

To assess luster correctly is purely a matter of experience; it will be relatively easy to recognize metallic and non-metallic luster by comparison with ordinary metal objects. A glassy, pearly, greasy or silky luster is self-explanatory but it is difficult to define a diamond-like or adamantine luster. This is best memorized by examining pale varieties of diamond. Intermediate lusters, as resinous or adamantine-metallic, are also difficult to define. Massive and granular aggregates are usually dull in luster.

To test the streak a short scratch is made upon a plate or tile of unglazed porcelain (obtainable from mineral dealers) using a corner of the specimen and sometimes further smearing the streak by rubbing with a second plate. The color of its streak is far more characteristic of a mineral than its body color.

Hardness is usually determined on the Mohs' scale (Table 8) which consists of 10 steps. Any would-be collector should get himself specimens of minerals 1–9 (easily obtainable from any dealer) and practise with these. Each member of the scale may be scratched by the one above it with moderate pressure. For rough preliminary tests it is useful to remember that gypsum (2) may be fairly easily scratched with a fingernail, while an ordinary pen-knife scrapes minerals of hardness 4 and 5. Plate glass is easily scratched by quartz, and only just by feldspar. Also to be taken into account is the fact that fine-grained to massive aggregates may be harder than crystals of the same mineral (massive gypsum can hardly be scratched by the fingernail). On the other hand loose aggregates and weathered

TABLE 8 The 10 minerals of Mohs' scale of hardness

1	talc $Mg_3Si_4O_{10}(OH)$	6	feldspar $KAlSi_3O_8$
2	gypsum $CaSO_4.2H_2O$	7	quartz SiO_2
3	calcite $CaCO_3$	8	topaz $Al_2SiO_4(F, OH)$
4	fluorite CaF_2	9	corundum Al_2O_3
5	apatite $Ca_5(PO_4)_3(F, Cl)$	10	diamond C

PLATE 9   *Realgar and orpiment*
Above Massive realgar. Leon, Spain. *Center* Foliated orpiment. *Below* Prismatic, striated crystals of realgar. Schlema, Erzgebirge, Germany.

19 C 64

specimens may appear to be softer than they really are, so that it may be impossible to determine hardness with any certainty.

Having determined luster, streak and hardness one enters the appropriate page and line in the tables. The second column gives the *color of the mineral,* which may, unfortunately, not be very characteristic at all: most minerals occur in a variety of colors and numerous shades. There are several reasons for this. In otherwise colorless minerals the smallest traces of admixture may cause appreciable coloration. Such admixtures may be chemical, where traces of other elements enter into isomorphous replacements, or physical if small grains or very fine needles of another mineral are present as inclusions. Radioactive influences may affect colors considerably (fluorite: variety antozonite; smoky quartz). The metallic sulphides and oxides have relatively stable colors, but their surface is practically always slightly altered and they tend to iridescence. Their true color can only be seen on freshly fractured surfaces.

Minerals are also classified as *transparent, translucent,* and *opaque.* All minerals with high metallic luster are opaque, and some otherwise transparent ones appear to be opaque if they are very highly colored; extremely thin edges on splinters of the latter will still be transparent, or translucent at least, as can easily be proved by using a lens.

Cleavage, a useful diagnostic property in minerals (see p. 28 and Table 10), is their ability to split easily in certain directions as a result of a blow, or pressure from a penknife. The degree of perfection of cleavage and the shapes of cleavage fragments vary from mineral to mineral. Galena readily breaks up into small cubes after a few light blows: galena is therefore said to have cubic cleavage. With a little care rhomb-shaped prisms may be cleaved out of hornblende. Mica and gypsum may be split into extremely fine flakes using a knife. The use of cleavage has its limitations of course because no one is likely to break up well-developed crystals just to determine their cleavage, but even the best crystals may have a bruised corner that can be inspected through a lens. This is also partly true for *fracture,* which is less characteristic. It is usually possible, however, to

PLATE 10   *Halite (rock salt)*
Partially distorted cubes on marl. Wieliczka, Poland. In front, isolated cleavage fragments.

recognize both properties on naturally fractured surfaces, or use inferior material for testing.

To determine *specific gravity* is rather cumbersome. The value given in the tables indicates in grams the density of one c.c. of a mineral compared with one c.c. of water (1 gram). A measuring cylinder and a scale balance will be needed (obtainable from firms selling laboratory equipment). The measuring cylinder is filled with water to a certain mark, which is noted, and the mineral fragment to be determined is put in. The new level of the water is read off. The difference between the two readings gives the volume of the fragment, which has previously been weighed. If the weight of the fragment is divided by its volume, its specific gravity is obtained. Example: A mineral fragment weighs 11.4 gms. and its volume is 4.3 c.c. 11.4 divided by 4.3 equals 2.65 (approx.), a value which indicates that the mineral may be quartz. This method is not very accurate and may only be used for single crystals or pure material. With practice on known substances it is possible to develop dexterity in guessing weights (compare calcite and barite).

The next two columns of the tables concern external shape. If one has crystals to deal with, one tries to determine the crystal shape with the help of Chapter 1. The remaining columns give information about the mode of formation, general distribution, and the most important associated minerals. Should the tables have helped you to arrive at a tentative identification, the main description of the mineral should now be consulted, together with any illustrations or diagrams.

Sometimes simple flame tests may also be made, using the flame of a blow-lamp. Some minerals fluoresce under ultra-violet light. A number of minerals cannot be identified at all without chemical, optical, or other determinations but this is always mentioned in the description. None of these minerals is included in the determinative tables, neither are those whose external characteristics cannot readily be determined because they only occur in tiny grains.

In conclusion it must be pointed out that use of the tables may not lead to a unique identification: many minerals are far too complex for this. Only experience will help, but even an expert may not always be able to distinguish individual members of the garnet or amphibole group, for example.

Part Two DESCRIPTION OF IMPORTANT AND COMMON MINERALS

About two thousand minerals are known, but of these only about a hundred are common and widespread, mainly the rock-forming minerals and the chief ores. These are described here. In addition a few rare but important minerals (diamond, pitch-blende) have also been included, and most of these are discussed in detail. Some others are only mentioned very briefly or appear only in the tables in Part Three.

The primary classification of minerals is based on chemistry and the atomic structure of the minerals. With the exception of the elements, subdivision is according to the anions of compounds in 8 classes:

1 Elements
2 Sulphides (S, As)
3 Halides (Cl, F etc.)
4 Oxides and hydroxides (O, OH)
5 Carbonates, borates etc. (CO_3, BO_3 etc.)
6 Sulphates and related compounds (SO_4, WO_4, MoO_4)
7 Phosphates and related compounds (PO_4, AsO_4, VO_4)
8 Silicates (SiO_4 etc.)

Each class is further subdivided, partly by detailed chemical composition and partly by structure.

In the historical notes the following names recur frequently: *Theophrastos,* Aristotle's pupil, author of the first book on stones, written about 320 BC; *Pliny the Elder:* victim of the eruption of Vesuvius in AD 79, author of a Natural History giving very detailed descriptions of minerals, ores, gemstones, rocks, and mining in ancient times; *Aristotle's* 'De Petris' was very probably written by an early medieval Arab scholar and – in keeping with the customs of those days – attributed to Aristotle; *Georgius Agricola* (1494–1555) summarized sixteenth-century knowledge of minerals, gemstones, mining, and quarrying; *Abraham Gottlob Werner* (1749–1817), mineralogist of Freiberg, Saxony, systematically described the external characteristics of minerals; *Martin Heinrich Klaproth* (1743–1817), discovered a number of chemical elements, and examined minerals in the light of their chemical compositions.

Abbreviations and symbols
  H = hardness according to Mohs' scale
SG = specific gravity
  (!) = occurrence of particularly fine specimens

Class 1 Native elements

Only a few elements are *native,* that is, occurring naturally in the pure or nearly pure state; only graphite (C) and sulphur (S) are common.

COPPER   Plate 1

CHEMISTRY Cu; usually contains small amounts of iron, gold; often mechanically admixed with silver.
STREAK bright metallic   CUBIC   H $2\frac{1}{2}$ – 3   SG 8.9
COLOR bright copper-red on fresh crystals, masses or scratches; often tarnished bronze or sometimes green.
LUSTER metallic to dull.
PROPERTIES no cleavage; fracture irregular; flexible, ductile, malleable.
CRYSTALS good crystals rare (cubes, octahedra, dodecahedra); mostly distorted crystals, skeletal, dendritic, wiry, lumpy, thin plates.
IDENTIFICATION reddish color, streak, softness, heaviness, malleability, flexibility; looks like copper!

The word 'copper' is derived from the island of Cyprus (Greek *Kypros)*. Native copper, the first metal used, was cold-hammered into various utensils in Egypt from 4500 BC.

  Copper is still one of the most-mined metals, after iron; the electrical industry is its chief consumer. Of its alloys bronze is still of importance but even more so are brass (compare tin, p. 140) and special alloys with aluminum, beryllium, etc.

PLATE 11   *Fluorite*
Above left Fluorite: the cube in the middle shows the 'parquet' structure well. Wölsendorf, Bavaria. *Above right* Clear green cubes of fluorite. Kinzigtal, Black Forest. *Below* Purple cubes of fluorite with octahedral cleavage cracks. Kinzigtal, Black Forest; all localities in Germany.

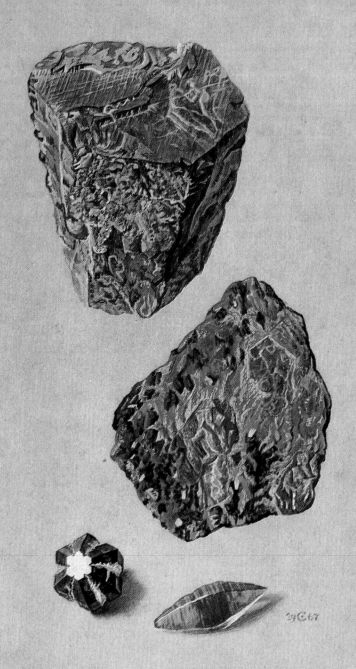

Chemicals containing copper are widely used in crop spraying and pest control.

Formation and occurrence. Native copper is mainly formed at the junction of zones of oxidation and cementation from copper sulphides, and also in the cavities of certain basaltic lavas. USA – copper mines of Keeweenaw Peninsula, Michigan, crystals (!!!) to several cm. width and in calcite (!!!), datolite, apophyllite, analcime, also nodules, hollow 'skulls', branches, dendrites, and masses (to several tons). In various copper mines of New Mexico and Arizona, abundant as branching growths (!) sometimes crystals (!). Rarely in zinc ore at Franklin, N.J., with calcite, zincite, willemite, etc. USSR – Bogoslovsk, Turinsk. GERMANY – Füsseberg, Siegerland, N. Rhine-Westphalia; Reichenbach, Odenwald. SOUTH-WEST AFRICA – in copper mines of Tsumeb and elsewhere. BOLIVIA – unique copper replacements of twinned aragonite crystals (!!!) to 2 cm. from Corocoro.

SILVER   Plate 1

CHEMISTRY Ag; often contains some gold, copper, mercury.

STREAK brilliant shining white   CUBIC   H 2$\frac{1}{2}$–3   SG 9.6–12

COLOR silver-white, often tarnished yellowish, brown, grey to black

LUSTER bright metallic.

PROPERTIES no cleavage; irregular fracture; malleable, etc., like copper.

CRYSTALS rare (cubes, octahedra, dodecahedra); twinned growths; wires, ropes, plates, dendrites, rounded masses, scales.

IDENTIFICATION color, heaviness, malleability, wire form.

Beads made of silver from Egypt date from before 4000 BC. In ancient times considerable silver was mined in the south of

PLATE 12   *Corundum and chrysoberyl*
Above Fragments of barrel-shaped corundum crystal. The horizontal surface and the striations indicate parting parallel to twin lamellae. Renfrew, Ontario, Canada. *Center* Ruby in pale-green massive zoisite with greyish-black hornblende. Longido, Tanzania. *Below left* Interpenetrant twinned crystal of alexandrite, a variety of chrysoberyl. Fort Victoria, Rhodesia. *Below right* Crystal of sapphire with curved faces. Ceylon.

Spain, but most came from the lead-zinc deposits of Laurion, Greece, which contributed to the political importance of Themistocles' Athens (sixth century BC). In late medieval times the great demand for silver caused mining concerns to proliferate in Central and Southern Europe, working the many lodes in the Harz Mountains, the Erzgebirge, and in the Tirol at Schwatz and Brixlegg.

Silver is one of the coinage metals, and is used for jewelry and very widely in industry, e.g. photographic chemicals.

Formation and occurrence. Occurs in oxidized portions of hydrothermal sulphide veins, also in cementation zones; hydrothermal silver formed in large quantities as at Kongsberg, Norway; plentiful in sulphide-carbonate veins in diabase in Ontario, Canada; appreciable quantities, with copper, Keeweenaw Peninsula copper mines, Michigan.

NORWAY – finest specimens from Kongsberg, in crystals (!!!), wires, ropes (!!!), etc., with calcite and silver sulphides (some ropy masses over 25 cm.). GERMANY – lodes at St Andreasberg, Harz; Freiberg, Schnessberg, Aue, etc., Saxony. CZECHOSLOVAKIA – Jachymov (Joachimstal). USA – crystals (!!), branching masses (!!) in copper mines Keeweenaw Peninsula, Michigan; micro crystals (!!) Creede, Colorado. CANADA – Cobalt, Gowganda, O'Brien, etc., Ontario, in large platy masses (up to 800 kg.) with sulphides in calcite veins; also dendritic. MEXICO – dendrites and 'herringbone' twin growths (!!) Arizpe, Sonora and especially Batopilas, Chihuahua.

GOLD   Plate 1

CHEMISTRY Au; only very rarely pure, usually with silver, a little copper, also platinum metals.

STREAK brilliant shining yellow   CUBIC   H $2\frac{1}{2}$ – 3   SG 15.5–19.5

COLOR golden yellow (brassy) to nearly silvery when containing much silver as in *electrum,* a variety containing more than 30 per cent silver.

LUSTER brilliant metallic.

PROPERTIES no cleavage; irregular fracture; very malleable, ductile, flexible.

CRYSTALS commonly small, rarely large, as complex distorted octahedra, sometimes showing good octahedral shape;

plates in cavities are mostly flattened octahedra; also wiry, nodular, branching, spongy, etc.

IDENTIFICATION non-tarnishing gold color, malleability, extreme heaviness; pyrite and other yellow metallic minerals are brittle; small flakes of mica confused with flake gold but are not malleable and very low SG.

The word 'gold' is related to *gelb* ('yellow' in German), or to Old High German *ghel*, 'shiny'. Electrum was named by Pliny, after the Greek *electron,* amber.

Man's first discovery of gold remains for ever shrouded in the obscurity of prehistoric times. The oldest Egyptian gold ornaments seem to be of slightly later date than the first copper objects. Gold was first obtained from gravels and later from mines. It is thought that prospectors from Crete had taken up gold mining in the Carpathians before 2000 BC. In Roman times gold was brought to Rome from Transylvania, from the Tauerns in Austria and from NW. Spain (6½ tons a year, according to Pliny).

The first 'gold rush' was in California in 1849. The gravels there yielded 70–100 tons yearly and laid the foundations for the economic wealth of the USA. The Victorian gold rush of 1851–52 trebled the population of Australia. The last 'rushes' were in 1896 near Klondike, in the Yukon, and in 1898 along the Yukon River, near Fairbanks, Alaska. The most notable occurrence today is the enormous deposit at Witwatersrand near Johannesburg, South Africa, first discovered in 1884, consisting of gold-bearing conglomerates which are probably compacted gravels. For some time Witwatersrand yielded 50 per cent of world production, totalling 20,000 tons between 1886 and 1960. The total amount of gold ever produced is estimated to be between 50,000 and 60,000 tons.

Formation and occurrence. Predominantly in hydrothermal quartz-sulphide veins in schists and slates, and alluvial deposits derived from their weathering; the very important deposit at the Witwatersrand contains very fine gold in a strong conglomerate rock.

USA – Crystals (!!!), dendritic-wiry masses (!!) in cavities with quartz crystals in mines of Mother Lode country, California, particularly at Grass Valley; skeletal crystals to 2 cm., also flattened octahedral plates to 25 cm. broad. Small sharp crystals (!!) Breckenridge district, Summit Co., Colorado, also from Red Mountain Pass, San Juan Co. ROMANIA – complexly crystallized fern-like growths on drusy quartz (!!!) from Verespatak, Bihar Mountains and elsewhere in Romania and the Carpathian Mountains. USSR – much alluvial gold in Urals and Siberia. GERMANY – 'Rhine gold' washed from alluvials of Rhine River since Roman times. AUSTRALIA – large crude crystals (!) and small sharper crystals (!!) from Bendigo and Ballarat, Victoria; also wire gold (!) from Flinders Range, South Australia; large nuggets have also been found, e.g. the 'Welcome' from Ballarat of 68 kg.

PLATINUM   Plate 1

CHEMISTRY Pt; usually alloyed with iron (up to 30 per cent) and the Pt-group metals osmium, iridium, palladium.
STREAK shining grey-white   CUBIC H 4–4$\frac{1}{2}$ SG 14–19, 21.5 pure
COLOR bluish-grey, like steel.
LUSTER metallic.
PROPERTIES no cleavage; irregular fracture; malleable, ductile, tough.
CRYSTALS very rare, mostly in stream-worn nodules ('nuggets').
IDENTIFICATION color, luster, malleability; very heavy.

Platinum was recognized as a distinct metal in 1750 and named 'platina' (diminutive of silver). It is obtained from gravels as well as primary deposits.

USSR – at Nizhni Tagil in the Urals, platinum occurs in dunites and gravels. COLOMBIA – at Choco, in gravels. Because of its high melting point (1774°C) and resistance to chemicals, laboratory crucibles are made from it, and it is also used in the electrical industry and in jewelry.

CARBON

Nowhere is the tendency of certain substances to occur in different modifications more marked than in carbon, with diamond and graphite as its two crystalline forms. Diamond consists of carbon tetrahedra, whereas in graphite the carbon atoms are arranged in sheets of hexagons. This accounts for their widely different crystalline forms and properties.

GRAPHITE   Plate 1

CHEMISTRY C; always more or less impure, containing quartz, clays, etc.

STREAK shining grey ('lead' pencil mark)   HEXAGONAL   H 1 SG 2.2

COLOR dark to medium grey.

LUSTER submetallic to dull.

PROPERTIES perfect cleavage; flexible flakes; greasy feel; easily marks paper.

CRYSTALS rarely well-formed; mostly in coarse to fine flaky masses.

IDENTIFICATION color, luster, ability to mark paper; distinguished from molybdenite because latter is much brighter luster with distinct silvery-bluish color.

Graphite was named by Werner from the Greek *graphein*, to write. In the seventeenth century graphite was mined as 'black lead' at Borrowdale, Cumberland, and used in lead pencils. Until the end of the eighteenth century the true properties of graphite were not known exactly. Its use in pencils is important but accounts for very little graphite only. Flaky graphite is compressed and made into crucibles (heat resistant to 3000°C) and is used also as a lubricant. Purest graphite is used in reactors, to keep the nuclear fission process under control by trapping the escaping neutrons.

There are large deposits in Korea and Austria (Kaiserberg near Leoben, Sunk near Trieben, both in Styria). Graphite there replaces coke in high-temperature furnaces. World production is about 300,000 tons a year.

Formation and occurrence. Graphite, formed from coal, is common in crystalline slates and marbles. The coarsely crystalline variety occurs in pegmatites together with quartz and feldspar. Korea and Austria have already been mentioned. GERMANY – Pfaffenreuth near Passau, Bavaria, in gneiss and in marbles of the area. CEYLON – Ragedara near Galle in pegmatites, formerly economically important. USA – Orange Co., and Ticonderoga, Essex Co., N.Y., occasionally in crystals.

DIAMOND   Plate 1

CHEMISTRY C; colorless nearly pure, but often containing

mechanical inclusions of other minerals, liquids, gases, or even other diamond crystals; also contains nitrogen in some types.

STREAK too hard to show on normal testing plates   CUBIC
H 10   SG 3.50–3.53

COLOR predominantly colorless with faint tinges of brown or yellow; also bright yellow, brown, black, green, red, blue, etc.

LUSTER adamantine, like faintly silvered glass.

PROPERTIES perfect octahedral cleavage, but fracture irregular in massive types with cleavage seldom visible; brittle in crystals, tough in massive types ('bort', 'carbonado').

CRYSTALS sharp octahedrons common, also cubes, dodeca-hedrons and rounded crystals of more complex forms; cubes and dodecahedrons often dull and pitted; also massive.

IDENTIFICATION crystal shapes, luster, supreme hardness (easily scratches corundum, glass, etc.).

In ancient times diamond was very rare, but it was known to be the hardest of all substances; the Greeks called it *adamas,* 'invincible', though there was a legend that it could be softened in fresh goat's blood. Pliny does not mention the cutting and polishing of diamonds; it is therefore likely that cut stones were introduced from India.

Ninety-five per cent (by value) of precious stones produced each year are diamonds. The rough stone is not at all striking to look at. Its 'fire', evoked by high luster, reflection and dispersion of light, is made apparent only by cutting. The 'brilliant' cut is the most effective of all, in which the angles, number and shape of the faces follow strict geometric rules.

All crystals unsuitable for cutting, as well as carbonados and borts are used as industrial diamonds, in rock drills, rock-cutting disks and saws, and for grinding and polishing ma-terials. Since 1955, industrial diamonds have also been made synthetically.

The weights of rough as well as of cut stones are given in carats (ct.), 5 to a gram. Present world production is about 36 million carats a year (7,200 kg., or rather over 7 tons), 80 per cent of which is for industry. Only a small proportion of the residue is of sufficiently high quality for gems, and over half of this material is lost in cutting. Few stones are over one carat, partly accounting for the high price, which also depends on color. Most highly priced are stones with a very pale blue coloration, referred to as 'blue-white'. Most stones are somewhat yellowish, but other colors exist and only about 10 per cent are colorless. Strongly colored stones are called 'fancy' diamonds.

The largest diamond ever found was the Cullinan (Premier mine, 1905), of 3,106 ct., or 621.2 grams; from it were cut two stones of 530 and 317 carats, as well as over 100 smaller stones, all of which are now among the British Crown jewels.

Formation and occurrence. Kimberlite, a peculiar basic rock, consisting of olivine (often altered to serpentine), chrome diopside, phlogopite, pyrope garnet, magnetite, hematite, ilmenite etc., occasionally contains diamonds. From the uniformly well-developed crystals, diamonds are believed to have been formed early from still liquid magma. These kimberlites form volcanic stacks, called 'pipes', and the diamond-bearing alluvial deposits are the result of the weathering of these pipes.

INDIA – the only known locality from ancient times until 1727, but no longer of importance. BRAZIL – Minas Gerais and Bahia (carbonados), discovered in 1727 and until 1900 the sole producer; compacted gravels. SOUTH AFRICA – in 1866 a young boy found a 21.75 carat diamond at the Orange River, near Hopetown. This led to the discovery of diamond gravels and in 1870 to the first pipes, from which the parent-rock of diamond became known for the first time. Very few pipes are worth working. The Premier mine produces 1 million carats a year. Alluvial deposits: the Vaal River, the coast in the vicinity of the southern bank of the mouth of the Orange River etc. SOUTH-WEST AFRICA – coastal gravels, now also worked from the sea with the help of dredgers. CONGO (Kinshasa) – the Kasai River produces nearly 60 per cent of the world's output, almost exclusively industrial diamonds. USA – single finds from the goldfields of California; also from Murfreesboro, Pike Co., Arkansas, in kimberlite. USSR – since 1949, gravels and pipes in Yakutia, Eastern Siberia have yielded much diamond.

Small diamond crystals, seldom over a carat and not of gem quality, may be purchased by collectors; also very small crystals displaying good forms as well as twins are available for micro-mounting at modest cost.

SULPHUR   Plates 2 and 28

CHEMISTRY S; nearly pure but sometimes contains selenium, tellurium, arsenic, or mechanical inclusions of clay, bitumen, etc.

STREAK white or colorless   ORTHORHOMBIC   H 2   SG 2

COLOR vivid yellow, orange-yellow; also reddish, greyish-brown.

LUSTER resinous to greasy, or vitreous.

PROPERTIES imperfect cleavage; extremely brittle; burns (!); transparent to translucent, sometimes opaque.

CRYSTALS often fine sharp bipyramids and ball-like crystals; also massive to somewhat porous earthy.

IDENTIFICATION vivid color; brittleness; lightness; combustibility.

Homer mentions sulphur as incense. According to Pliny, sulphur was mined on the Lipari Islands and near Naples, and used in medicine. Aristotle describes the healing qualities of sulphur springs, with stories about sulphur's miraculous qualities. To the alchemists, sulphur was the symbol of everything inflammable.

Sulphur is mainly used for making sulphuric acid, for the vulcanization of rubber, for making other chemicals, and in the manufacture of fireworks.

The extensive sulphur deposits of Louisiana and Texas, overlying saltdomes, are of the greatest economic importance. Water vapour at 163°C is pumped into them through bore-holes under high pressure, forcing molten sulphur upwards.

Formation and occurrence. Sedimentary sulphur is formed by the activity of sulphate-reducing bacteria. The largest deposits in marls and limestones containing aragonite and gypsum were

PLATE 13   *Hematite and magnetite*
Above Hematite, iridescent tarnished crystals. Elba, Italy. *Center left* Botryoidal, massive to finely fibrous hematite known as 'kidney ore'. Cumberland, England. *Center right* Magnetite octahedrons in chlorite schist. Zillertal Alps, Austrian Tirol. *Below* Hematite forming 'iron roses', also from the Zillertal Alps.

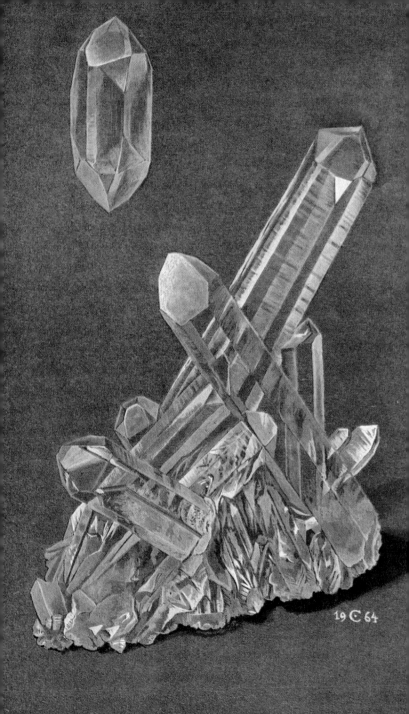

probably formed with the help of bituminous materials. Sulphur occurs associated with all active and extinct volcanoes and gypsum deposits.

ITALY – localities around Agrigento and Caltanisetta, Sicily, superb crystals (!!!), large and clear, with large aragonite twin crystals, calcite, celestite (!!) and gypsum; also Perticara near Cesena, Romagna; Sicilian crystals may reach 5 cm. diameter. USA – vast quantities in salt domes of Louisiana and Texas. MEXICO – small crystals (!) with gypsum, near San Felipe, Baja California; also volcanic deposits in the interior of Mexico.

Class 2 Sulphides

All sulphides are regarded as compounds consisting of one or more metals and of the non-metal sulphur. The semi-metal arsenic may substitute for sulphur, or either arsenic or antimony may be present together with sulphur; or, finally, arsenic or antimony may also take the place of the metal itself.

CHALCOCITE

CHEMISTRY Cu_2S, copper sulphide; sometimes with minor silver and other elements.

STREAK dark grey   HEXAGONAL (high chalcocite) or ORTHORHOMBIC (low chalcocite)   H $2\frac{1}{2}$–3   SG 5.6

COLOR dark grey to dead black.

LUSTER metallic when fresh, otherwise dull or 'sooty'.

PROPERTIES brittle; cleavage indistinct; opaque.

CRYSTALS uncommon, usually small tabular with hexagonal outline due to twinning; also massive, earthy; often pseudomorphous after other copper minerals.

IDENTIFICATION color; crystals; softness; resembles argentite but blacker and harder.

Chalcocite, from the Greek *chalkos,* copper, usually occurs close to the surface in the enrichment zone of copper deposits and it was therefore easily mined. It is the most important ore of copper.

PLATE 14   *Quartz: 1 – Rock crystal*
Above Perfect, doubly terminated crystal with unequally developed rhombohedrons and triangular trapezohedron faces (*cp.* Fig. 17, p. 131). Switzerland. *Below* Elongated prismatic crystals (*cp.* Fig. 19, p. 133). Madagascar.

Formation and occurrence. Two types of deposit are known to occur: hyrdothermal veins and impregnations; and deposits formed by the alteration and subsequent enrichment of other copper sulphides. The majority of chalcocites, the so-called 'disseminated copper ores', all belong to the latter group. Sedimentary deposits may sometimes occur, as in the copper slates of Mansfeld, Germany. Associated minerals are bornite, covellite, pyrite; other copper minerals, molybdenite, etc.

USA – fine sharp crystals (!!!) near Bristol, Connecticut to 2 cm. diameter; massive in large quantities at Butte, Montana and at Miami, Morenci, Bisbee and Ray in Arizona; also at Kennicott, Alaska. ENGLAND – in crystals (!!!) on matrix to 1 cm. from Redruth, St Just, St Ives, and Camborne in Cornwall. GERMANY – as replacements of fossil wood near Frankenberg, Hesse. ITALY – Calabona, near Alghero, Sardinia. TRANSVAAL – crystals (!!) at Messina.

BORNITE

CHEMISTRY Cu_5FeS_4, copper iron sulphide, variable copper content, commonly contains a little silver.

STREAK greyish-black   TETRAGONAL   H 3   SG 5.1

COLOR reddish-brown fresh, but tarnishing rapidly to purplish iridescent colors.

LUSTER sub-metallic.

PROPERTIES no cleavage, fracture uneven to subconchoidal, brittle, opaque.

CRYSTALS very rare as crude cubes, dodecahedrons; mostly massive.

IDENTIFICATION distinctive purplish tarnish and bronze color of fresh fracture surfaces; heaviness; softness.

Bornite, named after the Austrian mineralogist Ignaz von Born (1742–91), is an important copper ore.

Formation and occurrence. In hydrothermal ore deposits with other sulphides; minor quantities with prehnite and zeolites in cavities in basalt. USA – with prehnite in nodules and large crystals (!) Centreville, Fairfax Co., Virginia; Butte, Silverbow Co., Montana. ENGLAND – Redruth, Cornwall, crystals. AUSTRIA – crystals (!!) in cavities of Alpine type around the Gross Venediger. SWITZERLAND – Riffelberg near Zermatt.

ARGENTITE

CHEMISTRY Ag_2S, silver sulphide.
STREAK shining metallic black   CUBIC but low-temperature
   form is orthorhombic (?) ACANTHITE   H 2   SG 7.3
COLOR lead-grey to black.
LUSTER metallic to dull to black sooty.
PROPERTIES indistinct cleavages, subconchoidal fracture; not
   brittle, cutting into shavings like lead.
CRYSTALS commonly as rude cubes, sometimes partly hollow.
IDENTIFICATION resembles galena but much softer, no cleavage
   like latter, much less brittle, and 'shaves' with a knife tip.

The name is after *argent* for silver; an important silver ore.

Formation and occurrence. Primary argentite is formed in the
hydrothermal zone, usually in silver-bearing veins of galena.
Secondary argentite forms by enrichment between the oxidation and cementation zones. Argentite, being the silver-bearer
in galena, is almost always present as minute inclusions in
nearly all its occurrences (see galena, p. 87). As a primary
mineral it is associated with native silver and other silver
minerals, such as proustite, pyrargyrite; secondary association
with native silver, cerussite, chlorargyrite, etc.

   GERMANY – Freiberg, Saxony in crystals (!!!), also crystals
(!!!) from silver mines at Schneeberg, Annaberg, Marienberg;
crystals (!!) at Andreasberg, Harz. CZECHOSLOVAKIA – Jachymov, Kremnica, Banská Štiavnica in hydrothermal veins.
NORWAY – crystals (!!) with native silver, Kongsberg. MEXICO
– crystals (!) Arizpe, Sonora, and other silver mines in Guanajuato and Chihuahua.

PENTLANDITE

$(Fe, Ni)_9S_8$, cubic. Intimately intergrown with closely similar
pyrrhotite, from which it can hardly be distinguished without
a polished section. Most important nickel ore: Sudbury,
Canada; Pechenga, Russia (formerly Petsamo, Finland). For
formation see p. 44.

   Cronstedt discovered nickel in 1751 in niccolite. The miners
had already noticed this reddish mineral long ago, but since it
did not fulfil their hope of it being copper, and was even a
nuisance in smelting, it was regarded as a joke of the mountain

imps, the nickels, and named 'coppernickel'. Not until the nineteenth century did its mining begin, first for making new-silver (an alloy containing 10–70 per cent copper, 10–20 per cent nickel, 5–30 per cent zinc). Today most of the nickel is used in making various steels, and alloys, and for nickel-plating; a small portion for coins and chemicals.

SPHALERITE Plate 3

CHEMISTRY ZnS, zinc sulphide; up to 20 per cent Fe, also some Cd, Mn replacing Zn; Cu, Ag, Pb, Au commonly present.

STREAK pale yellow to brown ISOMETRIC H $3\frac{1}{2}$–4 SG 4

COLOR Dark to pale brown, black (Fe-rich var. *marmatite*), dark to pale green, red-brown, brownish-red, orange, yellow; sometimes nearly colorless.

LUSTER adamantine; earthy varieties dull.

PROPERTIES perfect, easy dodecahedral cleavage; brittle; transparent to opaque (*marmatite*); sometimes fluorescent UV.

CRYSTALS commonly as rounded complex tetrahedral-dodecahedral with twinning, seldom as simple easily-recognizable crystals; very commonly granular massive; also earthy, botryoidal ('Schalenblende').

IDENTIFICATION heaviness, cleavage, luster, color, translucency; rotten egg odor (HS) when immersed in acids; associated with metallic luster sulphides as galena, pyrite, chalcopyrite, etc.

Sphalerite is sometimes called blende, or zinc blende. Iron-rich varieties are known as marmatite, reddish ones as ruby-blende, massive banded ones as schalenblende, usually intergrown with galena (see Pl. 3). Sphalerite is by far the most important ore of zinc, but not until the nineteenth century could it be success-fully smelted; before then, smithsonite was the only usable ore.

Zinc became known in Europe during the sixteenth and seventeenth centuries, at first being imported from India and the Far East. Aristotle reports that the Mossinoiks by the Black

PLATE 15 *Quartz: 2 – Colored varieties*
Above Amethyst, part of a geode. Uruguay. *Center* Massive rose quartz. South-West Africa. *Below* Smoky quartz. Val Giuv, Switzerland.

19 © 64

19 C 64

Sea produced 'yellow copper' by smelting copper and smithsonite together – thus was brass discovered.

Large quantities of zinc are used today, especially for galvanizing steel and iron plates and in die-casting. One-fifth of the total production goes into making brass (10–40 per cent Zn, 60–90 per cent Cu).

Formation and occurrence. Most sphalerite is hydrothermal in origin (veins, replacements, etc.). Associated minerals, apart from ever-present galena, are: chalcopyrite, pyrite, pyrrhotite, arsenopyrite, also quartz, calcite, barite, fluorite.

ZnS also occurs in a hexagonal modification, known as wurtzite. It is not easily recognized and is rare. Some schalenblende is partly wurzite.

Sphalerite is abundant and easily obtained in fine crystal specimens, and only a few localities can be given. GERMANY – in crystals (!) from mines in N. Rhine-Westphalia as at Essen-Borbeck, Katzwinkel near Betzdorf, Ameise mine near Siegen, also in the Rhineland-Pfalz at the Holzappel mine between Bad Ems and Limburg, the Rosenburg mine near Braubach am Rhein, and others; good schalenblende from Stolberg in N. Rhine-Westphalia and from the famous ore deposits at Rammelsberg in the Harz. SWITZERLAND – crystals (!!!) in Alpine vugs in the Grimsel region, Tavesch, and exceptional crystals in the marble of Lengenbach, Binnatal. AUSTRIA – schalenblende at Bleiberg, Carinthia. YUGOSLAVIA – superb crystals (!!!) of shining black marmatite at Trepča. SPAIN – gem quality (!!) cleavages and crystal druses (!!) of transparent yellow, red, brown material from Picos de Europa, Santander. ENGLAND – crystals (!) from lead-zinc mines of Cumberland, Durham, and mines of Cornwall. USA – large crystals (!!) from the lead-zinc deposits around Joplin, Missouri. MEXICO – fine clear green crystals of gem quality from Cananea, Sonora; also

PLATE 16   *Quartz: 3 – Variations in form and color*

Above left Rose quartz with well-formed crystals. Minas Gerais, Brazil. *Above right* Scepter quartz. Mexico. *Center* Smoky quartz: distorted, parallel-intergrown crystals showing skeletal development. Switzerland. *Below left* Phantom quartz. Minas Gerais, Brazil. *Below center* Rock crystal group covered with a dusting of chlorite. Maderanertal, Switzerland. *Below right* Smoky quartz: several crystals are intergrown and slightly rotated towards each other, forming a twisted group or 'gwindel'. Val Giuv, Switzerland.

crystals (!!) from Naica, Chihuahua, and other sulphide ore mines.

CHALCOPYRITE Plates 4 and 5

CHEMISTRY $CuFeS_2$, copper iron sulphide.
STREAK greenish-black TETRAGONAL H $3\frac{1}{2}$-4 SG 4.2
COLOR bright brass to golden yellow, commonly black tarnished or iridescent.
LUSTER metallic.
PROPERTIES no cleavage, good conchoidal fracture, very brittle, opaque.
CRYSTALS usually small distorted tetrahedrons; also massive and coating other sulphide minerals.
IDENTIFICATION yellower than pyrite and more brittle; crystals distinctively pyramidal (tetrahedrons).

Agricola mentioned chalcopyrite in the sixteenth century. A reniform massive variety (rare) is sometimes known as blister copper ore. Chalcopyrite is the most widespread ore of copper but at the same time also the poorest.

Formation and occurrence. Chalcopyrite forms in veins under a variety of conditions. Occasionally basic plutonic rocks contain chalcopyrite. It rarely occurs in pegmatites but is common in pneumatolytic veins and replacements. Chalcopyrite is an important constituent in primary disseminated copper ores. Pyrite deposits and skarn ores practically always contain some chalcopyrite. Because of its wide distribution, only a very few localities can be mentioned.

Good crystal specimens abundant from MEXICO – La Bufa, Chihuahua, and Naica, among other sulphide deposits. USA – small sharp tetrahedrons on dolomite, around Joplin, Missouri in the lead-zinc mines. JAPAN – crystals (!!!) Ani and Arakawa, Ugo. BRITAIN 'blister ore' and crystals (!!) from various Cornwall mines; crystals (!!!) Greenside mine, Westmorland, also Wanlockhead mine, Dumfriesshire, Scotland. SWEDEN – the celebrated Kopparberg (Copper Mt.) at Falun has been worked since AD 1220. GERMANY – crystals (!!) Horshausen, Westerwald and Friedrich mine at Herdorf, near Betzdorf, N. Rhine-Westphalia, and also in many other sulphide deposits in the Harz and Saxony.

ENARGITE

CHEMISTRY Cu_3AsS_4, copper arsenic sulphide; some antimony
   may substitute for arsenic.

STREAK greyish-black   ORTHORHOMBIC   H 4   SG 4.4

COLOR grey-black to sooty black.

LUSTER submetallic to dull.

PROPERTIES perfect prismatic cleavage; fracture uneven;
   brittle; opaque.

CRYSTALS tabular, with blunt ends and striations along sides;
   cross-section narrow diamond-shaped; also granular massive.

IDENTIFICATION cleavage; crystals with blunt ends and
   striations.

Named from the Greek for 'visible', *enargos,* because of the
obvious cleavage. A minor ore of copper.

   Formation and occurrence. Medium-temperature hydrothermal
mineral in ore veins, associated with sulphides and quartz.
USA – crystals (!!!) at Leonard mine, Butte, Silver Bow Co.,
Montana; from Bingham, Tooele Co., Utah; small crystals (!)
at Ouray, Ouray Co., Colorado. PERU – brilliant crystals (!!)
Cerro de Pasco.

Tennantite tetrahedrite group   Plate 5

CHEMISTRY tennantite $Cu_{12}As_4S_{13}$, copper arsenic sulphide and
   tetrahedrite $Cu_{12}Sb_4S_{13}$, copper antimony sulphide, form a
   solid solution series.

STREAK reddish-brown to black   CUBIC   H 3–4   SG
   4.4–5.4

COLOR tetrahedrite steel-grey but iron-black when containing
   some iron and zinc; tennantite is bluish-grey, often tarnished
   to dull grey or black.

LUSTER sub-metallic.

PROPERTIES no cleavage, conchoidal fracture, brittle, opaque.

CRYSTALS tetrahedrons in tetrahedrite, often highly modified
   by other forms and sometimes coated with chalcopyrite;
   modified cubes in tennantite; commonly massive.

IDENTIFICATION crystal form, color, luster, softness.

84

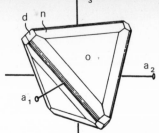

15 Tetrahedrite crystal. o, Tetrahedron. n, Tristetrahedron. d, Dodecahedron. Other steeper tristetrahedrons truncate the edges (striations).

Tennantite was named in honor of the English chemist Smithson Tennant (1761–1815).

Formation and occurrence. Tennantite-tetrahedrite is formed predominantly in hydrothermal veins.

ENGLAND – chalcopyrite-coated tetrahedrite crystals (!!) from Herodsfoot mine, Liskeard, Cornwall; small crystals (!) tennantite from Trevisans, Cornwall. GERMANY – crystals (!!!) to 2 cm. on edge from Clausthal, Harz; also from Horshausen, Westerwald, in the Harz, Saxony, etc. AUSTRIA – large black crystals (!) from Schwatz and Brixlegg (var. 'schwatzite'). SWITZERLAND – tennantite (var. 'binnite') in crystals (!!!) of considerable complexity in the dolomitic marble of Lengenbach, Binnatal. CZECHOSLOVAKIA – Příbram. ROMANIA – Baiá Spriei, Capnic (Kapnik), etc. SOUTH-WEST AFRICA – micro crystals (!!) at Tsumeb. PERU – crystals (!!!) Cerro de Pasco and Morococha. MEXICO – large crystals (!!) Mina Bonanza, Concepcion del Oro, Zacatecas. USA – crystals (!!!) from Bingham, Utah; micro tennantite crystals from Butte, Montana.

PYRRHOTITE Plate 4

CHEMISTRY FeS, iron sulphide, usually with excess sulphur.
STREAK black   HEXAGONAL   H4   SG 4.6
COLOR dull grey-yellow to bronze.
LUSTER metallic.
PROPERTIES distinct platy cleavage in aggregates of crystals, also conchoidal fracture, brittle, opaque; distinctly magnetic.

PLATE 17 *Quartz, flint, etc.*

Above left Perfect, doubly terminated crystal of quartz. Suttrop, Westphalia, Germany. *Center* Nodule of flint in hornfels. Neuberg on the Danube, Germany. *Below left* Flint nodule with a white weathered surface. Straubing, Lower Bavaria, Germany. *Below center* Silicified wood. Patagonia. *Below right* Fragment of a flint nodule from the chalk of Rügen, Friesian Islands.

CRYSTALS uncommon as flat plates of hexagonal outline, rosettes of such plates, or tapered hexagonal prisms; mostly massive.

IDENTIFICATION color duller than pyrite or chalcopyrite, bronzy; distinct magnetism most useful test.

The name comes from the Greek *pyrrhos,* meaning 'reddish'.

Formation and occurrence. Four types of deposit are known: liquid magmatic, contact pneumatolytic, hydrothermal, metamorphic pyrite deposits. FeS in meteorites is called *troilite.*

YUGOSLAVIA – superb crystals (!!!) to 10 cm. diameter and in rosettes, clusters, etc., at Trepča, which deposit still being mined. ROMANIA – crystals (!!) at Herja mine, Kisbánya. CANADA – crystals (!!!) and large, from Riondell, British Columbia. MEXICO – fine crystals (!!) to 6 cm. diameter Santa Eulalia, Chihuahua.

NICCOLITE

CHEMISTRY NiAs, nickel arsenide.

STREAK pale brownish-black   HEXAGONAL   H $5\frac{1}{2}$ SG 7.7

COLOR very pale coppery red, tarnishing to black, sometimes developing alteration coating of green annabergite.

LUSTER metallic.

PROPERTIES uneven fracture, no cleavage, brittle, opaque.

CRYSTALS very rare, mostly massive.

IDENTIFICATION pink metallic hue.

Formation and occurrence. Almost entirely restricted to hydrothermal veins with Ni-Co arsenides and silver. CANADA – Cobalt, Ontario, large masses. GERMANY – veins at Kinzigtal, Black Forest; St Andreasberg, Harz, and in the Erzgebirge at Schneeberg, Annaberg and Johanngeorgenstadt. CZECHOSLOVAKIA – Jachymov (Joachimstal).

GALENA   Plate 6

CHEMISTRY PbS, lead sulphide, almost always containing appreciable and sometimes important quantities of silver, making it an ore of that metal.

PLATE 18   *Chalcedony, agate and opal*
Above Chalcedony. Hüttenberg, Carinthia, Austria. *Center* Agate: polished section with varicolored concentric bands. *Below* Gem opal. Australia.

STREAK grey-black   CUBIC   H $2\frac{1}{2}$ SG 7.5

COLOR lead-grey, dark grey, rarely brilliant silvery with bluish tinge; tarnishes to black.

LUSTER metallic to dull, brilliant on fresh cleavages.

PROPERTIES perfect, easily developed cubic cleavage and very rarely fractures, very brittle, opaque.

CRYSTALS very commonly in cubes, cubes with octahedral modifications, and rarely as octahedrons; seldom without jointed surfaces ('mosaic' structure), sometimes with rounded edges and faces; also massive.

IDENTIFICATION perfect cubic cleavage, crystals, color, heaviness.

Galena was the name used by Pliny. It is a very common mineral and the chief ore of lead. In ancient times, and more recently in the East, powdered galena ground up in oil was used as a cosmetic; Aristotle in his book on stones *(De Petris)* mentions it in this context as the 'Itmid' stone. A statue of St Barbara, the patron saint of mining, made from massive galena, was erected in the eighteenth century in a Polish convent.

Lead was known to the Egyptians by 3000 BC. The Greeks and Romans used it for water pipes, for fastenings in masonry, and for refuse bins. Spain and England were the chief producers of galena ores in Roman times. In England lead blocks bearing the names of Roman Emperors were found, dating from the first and second centuries AD, some with seals stating whether or not the silver had been extracted. Chief uses: electrical industry (accumulators, cable sheathing, paints, anti-knock agent in gasoline; alloys (type-metal), ammunition, and radiation shields.

Formation and occurrence. Galena is a typical hydrothermal mineral. It is invariably accompanied by sphalerite; other associated minerals are pyrite, chalcopyrite, and other copper and lead minerals. Galena almost always contains silver, usually as grains of other silver minerals or sometimes in solid solution, and is therefore an important silver ore; 40 per cent of the world's silver is obtained from the smelting of lead. Galena, together with sphalerite, is very widespread in metasomatic or replacement deposits; the most extensive lead or zinc deposits are of this type. Sedimentary formations are also known.

Fine galena crystals are abundant from many deposits and only a few notable specimen localities can be given. USA – abundant in the many lead-zinc ore deposits of the Mississippi River Valley, especially from the Tri-State area of Oklahoma-Kansas-Missouri as crystals (!!) of large size, sometimes in the rarer octahedral form (cubes to 15+ cm. across); lustrous cube-octahedrons from Leadville district, Lake Co., Colorado. MEXICO – brilliant bluish-silvery cube-octahedrons (!!!) from Naica, Chihuahua. BRITAIN – brilliant, complexly twinned crystals (!!!) from Truro, Liskeard, etc., Cornwall; crystals (!!) Alston Moor and Weardale mines; crystals (!!) Wanlockhead, Scotland. GERMANY – crystals (!–!!!) from many ore vein deposits in the Harz, Saxony, etc., also from Bensburg-Gladbach near Cologne; Ramsbeck, Sauerland; Bad Ems, Koblenz, Stolberg and Diepenlinchen near Aachen. AUSTRIA – Bleiberg, Carinthia. CZECHOSLOVAKIA – crystals (!–!!!) Příbram, Banská Stiavnica, etc. YUGOSLAVIA – Trepča, crystals (!!!). ITALY – Iglesias, Sardinia crystals (!!!)

CINNABAR   Plate 7

CHEMISTRY HgS, mercury sulphide.

STREAK scarlet   HEXAGONAL (trigonal) H 2–2½   SG 8

COLOR deep red to blackish red, sometimes altered to nearly black on surface.

LUSTER adamantine to submetallic.

PROPERTIES perfect prismatic cleavage, conchoidal to uneven fracture, somewhat sectile, transparent to translucent.

CRYSTALS rare and usually small, mainly massive from granular to fine granular and earthy.

IDENTIFICATION red color, streak, softness, luster.

The name is of Persian origin and means 'dragon's blood'. Both cinnabar, the chief ore of mercury, and mercury itself have been known for a very long time. Cinnabar was used as a pigment in Neolithic times, and is found in cave paintings. Mercury itself was mentioned by Theophrastos. It was used from early times for extracting gold and silver by amalgamation. Today mercury is used for many purposes in the electrical industry, for chemicals, for explosives, and in physical apparatus. Mercury and its compounds are highly poisonous.

Formation and occurrence. Cinnabar is only deposited from low-temperature solutions. In hot spring deposits, as at Steamboat Springs, Nevada, cinnabar is formed at 80°C but it more commonly occurs impregnating and replacing rocks and filling fissures.

SPAIN – the famous Almaden deposit has been worked for over 2,000 years; in 100 BC about 4,500 tons of cinnabar were sent to Rome; mainly granular massive but also crystals (!!) to 1 cm on matrix. YUGOSLAVIA – crystals (!!) from Mount Avala near Belgrade; the Idria deposit has been mined since 1470. CHINA – exceptional twinned crystals (!!!) on drusy quartz and greatly in demand. USA – Cahill mine, Humboldt Co., Nevada crystals (!!) to 1.5 cm. MEXICO – crystals (!) on calcite from Charcas, San Luis Potosi.

COVELLITE

CHEMISTRY CuS, copper sulphide.
STREAK shining grey-black   HEXAGONAL   H 1½   SG 4.7
COLOR deep indigo blue, often purple iridescence.
LUSTER sub-metallic to dull.
PROPERTIES perfect, easy basal cleavage, brittle, somewhat sectile, opaque.
CRYSTALS thin platy of hexagonal outline or in foliate masses.
IDENTIFICATION distinctive color, softness, cleavage.

Covellite, named after the Italian mineralogist N. Covelli, was first discovered at the beginning of the nineteenth century by Freiesleben in the Mansfeld, Germany copper shales, and by Covelli in lava from Vesuvius.

Formation and occurrence. Covellite is formed as a result of weathering or alteration of copper sulphides, as chalcopyrite and chalcocite.

YUGOSLAVIA – at Bor. ITALY – Calabona mine, Alghero, Sardinia masses of large crystals (!!!) but often tarnished completely black. USA – large pure masses, capable of being cut into ornaments, at Kennecott, Alaska; crystals (!!!) and foliate masses in the mines of Butte, Montana.

STIBNITE   Plate 7

CHEMISTRY Sb_2S_3, antimony sulphide.

STREAK lead-grey   ORTHORHOMBIC   H 2   SG 4.7
COLOR lead-grey with dark or bluish tarnish, sometimes bluish-silvery.
LUSTER metallic to dull when tarnished.
PROPERTIES perfect cleavage, easily developed; brittle but somewhat sectile and flexible; irregular fracture; opaque.
CRYSTALS (Pl 7) long prismatic to needle-like, in criss-cross aggregates or radiating clusters; also massive.
IDENTIFICATION distinctive crystals, cleavage, color, heaviness; massive material tested with drop of potassium hydroxide solution which turns yellow and leaves a red stain.

'Stibnite' comes from *stibium,* the Latin name of the metal antimony, of which stibnite is the main ore. The derivation of 'antimony' is uncertain. According to legend, a monk named Valentinus noticed that antimony compounds hastened fattening in pigs. He intended to use this method on his fellow monks to make them look corpulent and contented, but unfortunately most of them died. Thereafter the metal was called *antimonachium,* 'against monks'. Antimony was in use five thousand years ago, as a forerunner of eye-shadow, to beautify the eyes. In medieval times antimony compounds were highly prized as medicines. Nowadays antimony, as a constituent of alloys, is used in such things as type-metal, ammunition, etc.

Formation and occurrence. Stibnite is typically hydrothermal, occurring in veins with other sulphide minerals, quartz, carbonates, etc.

JAPAN – magnificent prisms and blades (!!!), the world's largest and finest, in crystals to 60 cm. in the deposits of Ichinokawa, Saijo, Iyo Province, Shikoku Island. ROMANIA – radiating sprays and rosettes of thick needles (!!) on matrix from Felsöbánya (Baiá Spriei), Kisbánya, and Săcărămb (Nagyág), the sprays with crystals to 5–10 cm. long. YUGO-SLAVIA – very large crystals (!!) at Zajača; also at Stolice and Alšar. CZECHOSLOVAKIA – crystals (!) at Kremnica (Kremnitz). ITALY -- crystals (!) on matrix with realgar and cinnabar at Pereta, Tuscany. USA -- stout prisms (!!) at Manhattan mines, Nye Co., Nevada, the best North American specimens. Formerly fine crystals from Darwin, Inyo Co., California, also Rand district, San Barnardino Co.

The Japanese specimens are very difficult to obtain and are very costly; lately the market has been supplied with modest to fine specimens from Romania.

PYRITE   Plate 8

CHEMISTRY FeS_2, iron disulphide; nickel or cobalt may substitute for the iron; when more nickel than iron is present, the species bravoite, $(Ni, Fe) S_2$, is formed.

STREAK greenish or brownish black   CUBIC   H $6-6\frac{1}{2}$
SG 5.1

COLOR pale brassy yellow to golden yellow; sometimes iridescent or coated with brown limonite film.

LUSTER brilliant metallic.

PROPERTIES no cleavage, excellent conchoidal to irregular fracture; brittle; opaque.

CRYSTALS very commonly as cubes and pyritohedrons, uncommonly as octahedrons; also as modified individuals; cubes with characteristic striations (see Pl. 8, also Fig. 16); interpenetrant twins of two pyritohedrons are known as 'iron cross' twins. Also very common as granular, rounded, fibrous radiating nodules, and other massive forms; disseminated grains occur in almost any type of rock; alters to brown to dark brown limonite, preserving original crystal shape.

IDENTIFICATION harder than sulphide minerals which resemble it (e.g., chalcopyrite, pyrrhotite), while color less yellow than chalcopyrite and lighter than pyrrhotite but yellower than marcasite; unlike chalcopyrite and pyrrhotite, produces abundant sparks when struck against steel; scratches glass; crystals very typical.

'Pyrite' is derived from the Greek *pyr,* meaning fire, alluding to the sparks it gives off on friction. In Arabic it appears as *markasita.* In ancient times the term 'pyrites' covered copper and iron ores generally, as well as pyrite, and this caused a lot of confusion, especially with chalcopyrite.

Pyrite serves as a source of sulphur for the manufacture of sulphuric acid. Sulphur is converted to sulphur dioxide by 'roasting' and then turned into sulphuric acid. Metals other than iron are removed from the residual slag and then iron is extracted from the residue.

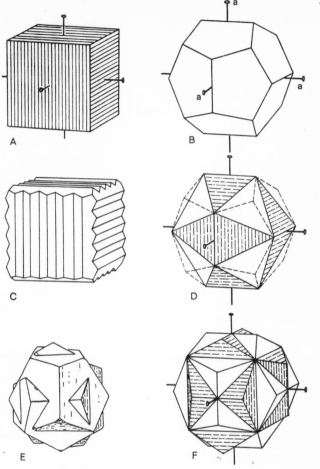

16 Pyrite crystals. (A) Cube with characteristic striations. The cube has lost the 4-fold symmetry shown in Fig. 2. (B) Pentagonal dodecahedron ('pyritohedron'). (C) Multiple combination of cube and pentagonal dodecahedron (striations on the cube face are caused by protruding edges of pentagonal dodecahedra). (D) Combination of pentagonal dodecahedron (shaded) and octahedron (unshaded). Both sets of faces are equal in size. Because of its similarity to the regular icosihedron this form is often called 'icosihedral combination'. The shaded areas complement the pentagonal dodecahedral faces of the figure shown in B. (E) Natural distorted 'iron cross' twinned crystal. (F) The same twinning reconstructed with one individual shaded.

Formation and occurrence. Pyrite, the most abundant sulphide mineral, is a perfect example of a vein mineral. Formed under many conditions, it occurs, perhaps, least of all in plutonic rocks, or in ore-bodies formed by liquid-magmatic segregation. Pyrite is common in pneumatolytic veins of all kinds and is invariably present in the various formations of the hydro-thermal zone. It also occurs in sediments of all types, disseminated throughout or sometimes sufficiently enriched to form deposits on its own, and it is a common constituent of coal. It may often form extensive deposits with or without chalcopyrite. It does not occur in the oxidation zone because it is decomposed there to limonite, resulting in the formation of the so-called 'iron-cap' or 'gossan'. Consequently it is also missing from alluvial deposits. Only the most important of its many occurrences are mentioned here, particularly those with exceptional crystals.

SPAIN – Rio Tinto, Huelva Province, the world largest deposit. USA – Bingham, Tooele County, Utah, in crystals of 12 cm. diameter; large crystals (!!) from Park City, Wasatch Co., Nevada; also large crystals (!!) from Leadville, Colorado; 'pyrite suns' (!!!), beautiful discoidal concretions in coal shales near Sparta, Randolph Co., Illinois; curved octahedral crystals (!) at French Creek mines, Chester Co., Pennsylvania; striated cubes in chlorite schists at Chester, Vermont. MEXICO – abundant crystals (!) from many lead-zinc-silver mines of Chihuahua and other states. ITALY – Agordo, Venetian Alps, deposit; from Brosso and Traversella, Piedmont, in pyrito-hedrons (!); Gavorrano, Tuscany, as fine crystals; Rio Marina, Isle of Elba, exceptional crystals (!!!) associated with hematite. GERMANY – Vlotho near Minden, Westphalia, limonite pseudo-morphs, interpenetrating 'iron cross' twins (!!) in marl; Münstertal, Black Forest, fine crystals; Bundenbach and Wissenbach, Rhineland, replacing fossils in shale; Nürtingen on the Neckar, pyritized ammonites. AUSTRIA – Habachtal,

PLATE 19 *Rutile and cassiterite*
Above Striated crystals of rutile, some of which are geniculated (knee-shaped) twins, with yellowish-white albite and quartz from an alpine fissure. Pfunds, Tirol, Austria. *Below left* Smoky quartz with inclusions of yellow needles of rutile in random orientation. Minas Gerais, Brazil. *Below right* Interpenetrant twinned crystals of cassiterite. Erzgebirge, Germany.

Hohe Tauern, crystals in slate; Waldenstein and Bleiberg-Kreuth, Carinthia, pyritized fossils. SWITZERLAND – in crystals (!!) from St Gotthard and Binnental, Valais, in dolomite marble. YUGOSLAVIA – Trepča and elsewhere, in pseudomorphs after pyrrohotite (!!); Majdan-Pek, pyrite deposit. CZECHOSLOVAKIA – Kremnica and Banská Stiavnica, Slovakia. ENGLAND – as pyritized ammonites (!!), Lyme Regis, Dorset; in fine crystals from Cornwall; from Cumberland with quartz and barite. USSR – Berezovsk, Ural Mts. (see gold, p. 68), crystals of 30 kg.

MARCASITE   Plate 8

Orthorhombic modification of FeS_2 (for derivation of the name, see pyrite). Of hydrothermal origin, it occurs also in sediments, like pyrite and together with it. Only well crystallized forms (spear-shaped, 'cock's comb') are easily recognized. Radiating nodules of marcasite found in collections are almost without exception pyrite.

In many hydrothermal veins in Germany. In Tertiary clays with lignite in Czechoslovakia; in Great Britain as spear-shaped twins from the chalk between Folkestone and Dover, Kent; cock's comb aggregates at Tavistock, Devon. Jewelers' 'marcasite' is pyrite. Large marcasite crystal groups (!!) are found with galena and sphalerite in the lead-zinc deposits around Joplin, Missouri.

COBALTITE

CHEMISTRY CoAsS, cobalt arsenic sulphide, always containing some iron.
STREAK grey-black   CUBIC   H $5\frac{1}{2}$ SG 6.2
COLOR tin-white; massive specimens greyish with slight pink hue.
LUSTER metallic.
PROPERTIES good cubic cleavage; uneven fracture; brittle; opaque.

PLATE 20   *Uranium minerals*
Above Uranophane in bluish-black fluorite (antozonite); the uranophane needles, irregular to radiating, are shown much enlarged in the circle (center). Wölsendorf, Bavaria, Germany. *Below* Massive pitchblende from Jachymov, Czechoslovakia.

CRYSTALS rare; usually cubes or pyritohedrons, or combinations.

IDENTIFICATION resembles pyrite crystals but tin-white color is distinctive; softer than pyrite; in massive forms a very pale pink hue distinguishes from smaltite and other sulphides; nitric acid (warm) becomes pink when powdered cobaltite is added.

Cobalt is derived from the German *Kobold* (goblin, elf); in the Middle Ages cobalt and nickel ores were regarded as having been produced by evil gnomes, as not a single scrap of copper or silver could be obtained from these ores, in spite of their appearance, and what is more they also gave off dangerous fumes of arsenic. Later, cobalt ores were highly prized for the manufacture of the pigment cobalt-blue. Cobalt is an important constituent of many alloys, special steels, etc., and is used to cement and toughen tungsten carbide for cutting tools. Cobaltite, in common with a number of other cobalt sulphides and arsenides (e.g. safflorite), is an important ore of cobalt.

Formation and occurrence. Cobaltite occurs widely in pneumatolytic ore deposits metamorphosed to skarns, associated with chalcopyrite, pyrrhotite, other sulphides, magnetite, garnet etc.; also in hydrothermal veins with other cobalt and nickel arsenides, niccolite, native silver, and argentite (acanthite).

CANADA – crystals (!!) to 2 cm. in mines near Cobalt, Coleman Township, Ontario. SWEDEN – the finest crystals (!!!) occur at Tunaberg, Riddarhyttan and elsewhere. Crystals are difficult to obtain and expensive.

LOELLINGITE

$FeAs_2$, orthorhombic. Was originally found in the Lölling district of Hüttenberg, Carinthia, as radiating or granular aggregates in siderite. Resembles arsenopyrite but crystals usually prismatic instead of stubby as in arsenopyrite. Good crystals have been found in the Langesund Fjord district, Norway.

ARSENOPYRITE

CHEMISTRY FeAsS, iron arsenide-sulphide, cobalt usually present.

STREAK black   MONOCLINIC   H $5\frac{1}{2}$–6   SG 6
COLOR tin-white to steel-grey.
LUSTER bright to dull metallic.
PROPERTIES distinct cleavage, uneven fracture, brittle; opaque.
CRYSTALS common, wedge-shaped, sometimes prismatic with rhombic cross-section; also massive granular.
IDENTIFICATION tin-white color, crystal form, hardness.

The name is derived from 'arsenic', which in turn came from the Greek *arsenikos,* masculine – the name Theophrastos gave to orpiment. The poisonous nature of arsenic has been known since ancient times. Arsenical compounds are widely used in pest control, and also in medicine (e.g. salvarsan); the metal is used in alloys (bearings, lead shot). Arsenic is widely obtained as a by-product, and sometimes causes disposal problems. At Boliden in Sweden, for instance, where arsenic is a by-product of the mining of copper, silver, gold etc., silos have had to be built for the storage of arsenical waste, to prevent spreading contamination.

Formation and occurrence. Primarily in metasomatic deposits (for paragenesis, see under cassiterite, p. 140), also in hydro-thermal veins (quartz-gold, silver and siderite).

GERMANY – crystals (!!) at Freiberg, Saxony, also crystals at Grube Bayerland, near Waldsassen, Bavaria. CZECHOSLOVAKIA – crystals from the tin lodes of Cinvald (Zinnwald), Ehren-friedersdorf, and Schlaggenwald. AUSTRIA – sharp crystals (!) in schist at the Mitterberg, near Muhlbach. ITALY – crystals from the Trentino. YUGOSLAVIA – splendant crystals (!!!) in large groups from Trepča and Alšar, near Dunje, Macedonia. ENGLAND – crystals (!!) at Tavistock and in Cornwall. MEXICO – crystals (!!!) Hidalgo de Parral, Chihuahua, and from Naica and other mines, abundant and easily obtainable from dealers. BOLIVIA – crystals (!!!) to 4 cm. diameter from Llallagua.

MOLYBDENITE
CHEMISTRY MoS_2, molybdenum disulphide.
STREAK blue-grey   HEXAGONAL   H 1–$1\frac{1}{2}$ SG 4.8
COLOR blue-grey.
LUSTER brilliant metallic.
PROPERTIES perfect, easy basal cleavage producing flakes, flexible but not elastic, opaque.

CRYSTALS hexagonal plates with tapered edges, usually small isolated or in larger curving or twisted masses; rarely sharp and distinct; also in small scattered flakes in matrix.

IDENTIFICATION somewhat resembles graphite but color distinct blue and luster very much more metallic; the color and luster resemble freshly broken galena; flaky cleavage; softness.

The name comes from the Greek *molybdos,* lead; for a long time it was suspected that molybdenite contained lead. The existence of molybdenum, a distinct metal, in this mineral was not discovered until 1782. Molybdenite is the chief ore of molybdenum, which is also obtained as a by-product from certain copper ores. It is much sought after for special steels. On account of its extremely good cleavage, molybdenite is also a valuable lubricant.

Formation and occurrence. Either pneumatolytic or hydrothermal; occurs in quartz veins with pyrite; veins associated with tin lodes; some hydrothermal disseminated ores; in skarns. Widespread in small quantities but concentrations are rare.

CANADA – large crystals (!!!) to 8 cm. diameter in Grenville marbles and pegmatites of Ontario and Quebec, especially fine at Aldfield, Pontiac Co., Quebec. USA – crystals (!!) to 5 cm. diameter in quartz, Lake Chelan, Washington. JAPAN – sharp crystals (!!!) from Hirase mine, Honshu. AUSTRALIA – superb crystals (!!!) in and with quartz crystals from Kingsgate and Deepwater districts, New South Wales. KOREA – notable crystals (!!) in various mines. Ordinary specimens are easily obtainable but sharp crystals are rare and expensive.

The skutterudite-safflorite group

Two groups of minerals are, in fact, discussed here together, but individual mineral members resemble each other so closely

PLATE 21   *Limonite and pyrolusite*
Above Botryoidal limonite with a shiny, lacquer-like surface. Herdorf, Germany. *Center* Radiating pyrolusite as pseudomorphs after manganite. *Below* Cellular-stalactitic limonite. Herdorf, Germany.

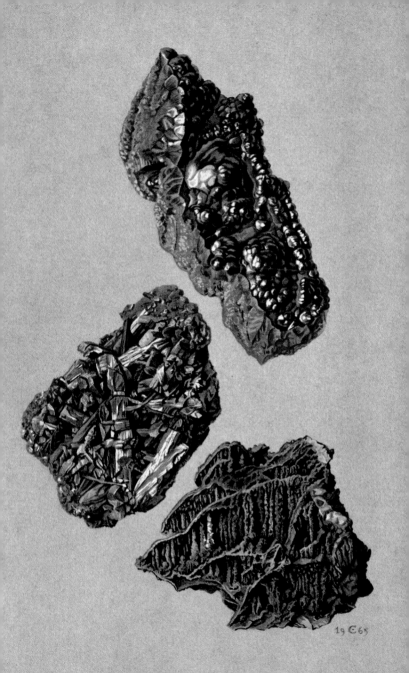

that it is impossible to distinguish them without resorting to chemistry, microscopy, or X-ray methods.

SKUTTERUDITE	$CoAs_{3-x}$	H $5\frac{1}{2}$–6	SG 6.6
CHLOANTHITE	$NiAs_{3-x}$	H $5\frac{1}{2}$–6	SG 6.6
SAFFLORITE	$CoAs_2$	H $4\frac{1}{2}$–5	SG 7.2
RAMMELSBERGITE	$NiAs_2$	H $5\frac{1}{2}$–6	SG 7.1

Complete solid solution series exist between skutterudite and chloanthite and between safflorite and rammelsbergite. Fe can always enter into it. Skutterudite (named after a place in Norway) is also called smaltite (from smalt, meaning cobalt-blue). 'Safflorite' is derived from the safflower, a thistle-like flower whose dried petals yield a red dye. All members of the group are tin-white, have bright metallic luster and black streak. The cobalt-rich members often appear to have a reddish coating of erythrite (hydrated cobalt arsenate), while nickel-rich members have a similar but green coating of annabergite (hydrated nickel arsenate; *chloanthes* means a green efflorescence). In massive specimens the minerals are closely intergrown with each other, and also with arsenopyrite, cobaltite, niccolite, loellingite and others. See localities given for these minerals.

Identification. Only the small cubic crystals of skutterudite can sometimes be identified; it is impossible to distinguish massive forms of these minerals by eye. When small fragments are dissolved in nitric-acid, Co-rich material will color the solution red, Ni-rich material green (be careful: corrosive fumes!).

Formation and occurrence. In hydrothermal ore veins with nickel and cobalt sulphides. CANADA – small but sharp crystals of smaltite at Cobalt, Ontario. NORWAY – crystals of skutterudite at Skutterud, near Modum. GERMANY – chloanthite Andreasberg, Harz and at Annaberg, Saxony; chloanthite (!!) crystals Schneeberg, Saxony. Crystals rare and good specimens expensive.

PLATE 22 *Calcite*
Above 'Butterfly' twin of calcite. Cumberland, England. *Center* Group of calcite crystals. Cumberland, England. *Below* Iceland spar. The transparent cleavage fragment is placed on a cross of two white lines to show double refraction. Helgustadir, Iceland.

THE RUBY SILVERS

PROUSTITE Ag_3AsS_3 (with a little Sb), and PYRARGYRITE Ag_3SbS_3 (with a little As), light and dark ruby silver respectively, have similar properties.

STREAK proustite 'vermilion, pyrargyrite cerise   HEXAGONAL
    H $2\frac{1}{2}$ SG 5.6–5.8

COLOR proustite vermilion to scarlet; pyrargyrite dark red to greyish-black. Both minerals blacken on exposure to light.

LUSTER adamantine, pyrargyrite also metallic.

PROPERTIES cleavage rhombohedral, brittle, transparent to translucent.

IDENTIFICATION distinguishable by streak.

The ruby silvers are valuable silver ores. Proustite was named after the French chemist J. L. Proust (1754–1826); pyrargyrite means 'fire silver' (Greek).

Formation and occurrence. Both minerals are limited to hydrothermal veins and occur together; proustite is slightly the rarer.

CHILE – superb crystals (!!!) of deep red transparent proustite from Dolores mine, Chanarcillo, the world's best, sometimes to 8 cm. long. CZECHOSLOVAKIA – crystals (!!) from Joachimsthal, Bohemia. GERMANY – crystals (!!) from Harz and Saxony mines (especially at Schneeberg). MEXICO – from various silver mines as at Guanajuato, etc., formerly fine specimens. Rare, attractive (proustite) and expensive!

THE FIBROUS SULPHOSALTS

This group includes a number of usually acicular-fibrous, or occasionally massive minerals. They can only be identified by chemical and similar methods. Examples are: JAMESONITE, Pb-Fe-Sb sulphide; PLAGIONITE and BOULANGERITE, Pb-Sb sulphides; JORDANITE, Pb-As sulphide. So-called 'PLUMOSITE' is largely boulangerite or jamesonite, but may also be stibnite. They occur mostly in hydrothermal veins, also in the dolomitic marbles of the Binnatal, Switzerland.

YUGOSLAVIA – excellent felted crystals (!!!) with sphalerite from Trepča (boulangerite). MEXICO – fine felted masses (!!) with pyrite from Mina Noche Buena, near Mazapil (jamesonite).

BOURNONITE

CHEMISTRY $PbCuSbS_3$, lead copper antimony sulphide.

STREAK grey   MONOCLINIC   H 3   SG 5.8

COLOR grey to black.

LUSTER brilliant metallic when fresh or unaltered, otherwise dull.

PROPERTIES poor cleavage, fracture subconchoidal, brittle, opaque.

CRYSTALS fairly common; usually as interpenetrating cruciform twins, sometimes with deep reentrant edges and forming the so-called 'cogwheel ores' of Cornish miners; also granular massive.

IDENTIFICATION recognized mostly on basis of typical crystals but massive kinds not easily distinguished from tetrahedrite–tennantite.

The mineral is named after the French mineralogist, Count J. L. de Bournon (1751–1825).

Formation and occurrence. It is primarily formed in Ag-Au veins, also in replacement deposits and in disseminated copper ores. Frequently accompanied by tennantite-tetrahedrite.

ENGLAND – the world's finest crystals (!!!), sharp and brilliant, on quartz matrix, have been taken from Herodsfoot mine, Cornwall; some crystals (twinned) reach 8 cm. diameter. GERMANY – fine crystals (!!) on matrix, Grube Georg, near Willroth, Rhineland–Pfalz; also from the Harz mines as crystals (!!). AUSTRIA – Hüttenberg, Carinthia, as crystals. USA – formerly fine crystals (!!) from Park City, Summit Co., Utah. MEXICO – small steel-grey crystals (!) from Naica, Chihuahua. BOLIVIA – large crystals (!!) to 10 cm. diameter from Vibora mine, Machacamarca. JAPAN – brilliant sharp crystals (!!) Nakaze mine, Hyogo Prefecture, Honshu. Prized by collectors but difficult to obtain and costly.

REALGAR   Plate 9

CHEMISTRY As_4S_4, arsenic sulphide.

STREAK orange   MONOCLINIC   H $1\frac{1}{2}$ SG 3.5

COLOR deep red.

LUSTER adamantine to resinous.

PROPERTIES distinct prismatic cleavage, conchoidal fracture, very brittle; transparent to translucent.

CRYSTALS small prismatic; mostly massive, coatings, veinlets.

IDENTIFICATION almost always associated with brilliant yellow orpiment; very soft and brittle; color; burns with garlic odor.

Realgar, a word used by the alchemists, comes from the Arabic. Natural realgar, as well as the synthetic compound, is used as a pigment, in the glass industry, and for fireworks.

Formation and occurrence. In low temperature hydrothermal deposits; hot springs deposits; nearly always associated with orpiment and commonly with calcite. USA – superb specimens, with orpiment, both in crystals (!!!) to 3 cm. at Getchell mine, Humboldt Co., Nevada; also crystals (!!) Mercur, Tooele Co., Utah. HUNGARY – crystals in clay at Tajowa, Neusohl. YUGO-SLAVIA – crystals (!!) Kresevo, Bosnia and at Alšar, Macedonia. TURKEY – crystals (!!) from Balin. The Getchell mine presently continues to supply fine specimens.

ORPIMENT   Plate 9

CHEMISTRY As_2S_3, arsenic sulphide.

STREAK bright yellow   MONOCLINIC   H $1\frac{1}{2}$–2   SG 3.5

COLOR vivid golden yellow, orange, to brown.

LUSTER adamantine to resinous; pearly on cleavage planes.

PROPERTIES perfect, easy cleavage; soft, sectile; transparent to translucent.

CRYSTALS usually poorly formed stubby individuals with rounded edges, dull luster; commonly massive foliate intimately associated with realgar.

IDENTIFICATION distinctive color; softness; sectility; cleavage; association with realgar.

Auripigmentum, mentioned by Pliny, which means 'golden-colored', has been corrupted to 'orpiment'. The mineral was known in the ninth century, and was, and in the East still is, used as a cosmetic. In China, orpiment has long been used for gilding silk. Arsenic is obtained from it now, and it is also used as a pigment.

Formation and occurrence. Formed with realgar, which see for occurrences.

Class 3 Halides

The halides are minerals that contain metals combined with fluorine, chlorine, bromine or iodine, i.e. the *halogens*. Only some fluorides and chlorides are common.

HALITE   Plate 10

CHEMISTRY NaCl, sodium chloride.

STREAK colorless   CUBIC   H 1   SG 2.1

COLOR colorless, also pale pink, orange, blue, purple, etc.

LUSTER glassy to oily.

PROPERTIES perfect and easy cubic cleavage; brittle; transparent to translucent.

CRYSTALS common, as cubes; sometimes skeletal; also granular massive, fibrous.

IDENTIFICATION crystal form, luster, and taste (sylvite, which looks like halite, tastes bitter); colors a flame yellow.

Halite is derived from the Greek *hals,* salt. The syllable *hall-* is often met in place names in the vicinity of salt deposits, such as Halle, Hallein, Hallstatt.

For thousands of years the sole use of salt was as a nutrient and preservative. At present only a minute fraction of the total production is for human consumption, although the average annual consumption is around 5.5 kg. per person. The chemical industry is the chief consumer (hydrochloric acid, soda, caustic soda, sodium metal).

Halite is quarried or mined, or produced by evaporation of sea water from either natural or artificial brine reservoirs, and from salt farms by the sea. Potassium salts are predominantly produced by quarrying but also from natural brine, as from the Dead Sea. Reserves are practically inexhaustible. The most common minerals of potash deposits are:

sylvite   KCl   colorless, white   salty, bitter

kainite $KMg (Cl/SO_4) . 2\frac{3}{4} H_2O$ white, yellowish, grey, red, purple   salty, bitter

carnallite $Kmg\, Cl_3 . 6H_2O$ red, white, yellow   sharp, bitter

kieserite $MgSO_4 . H_2O$ white, grey, yellowish, greenish
                                                tasteless

polyhalite $K_2Ca_2Mg(SO_4)_4 . 2H_2O$ brick-red, yellowish
                                                                   almost insoluble

Formation and occurrence. With the exception of a very little salt formed by volcanic sublimation, the formation of halite (and of the potassium salts as well) is entirely by evaporation of brine. The so-called Prairie Evaporites have formed from ground efflorescences in arid regions. Halite is always found in association with anhydrite and gypsum.

USSR – crystals (!!) from Iletskaya Zashchita, Siberia. GERMANY – fine blue crystals (!!!) from Stassfurt and Heringen am Werra, Hessen. POLAND – famous since 1100, the underground salt mines of Wieliczka and Bochnia near Cracow contain a chapel and other rooms, carved out of solid salt, with all furniture, altar, etc. also carved from the halite; beautiful crystals (!!!) from here. USA – recently pale orange fluorescent crystals (!) deposited from hot springs on twigs from near Salton Sea, Imperial Co., California; also from Searles Lake, San Bernardino Co. Good specimens inexpensive but due to tendency to take up atmospheric moisture, they are not popular and must be preserved in jars or plastic boxes.

FLUORITE   Plates 11, 20 and 47

CHEMISTRY CaF_2, calcium fluoride.
STREAK colorless   CUBIC   H 4   SG 3.2
COLOR colorless, commonly some shade of blue, purple, green; also pink, yellow, brown, nearly black, etc.; also color-zoned.
LUSTER glassy to somewhat oily.
PROPERTIES excellent octahedral cleavage, easily developed; rarely fractures conchoidally to irregularly; brittle; transparent to translucent; fluoresces under ultraviolet light.
CRYSTALS common as cubes but rarely sharply formed with smooth faces; mostly in 'mosaic' crystals with joints appearing on faces; considerably rarer as octahedrons; also granular massive, commonly color-banded.
IDENTIFICATION crystal form; cleavage most valuable clue; harder than halite, calcite.

The name fluorite alludes to the easy melting of this mineral and its use as a flux in the smelting of ores (from the Latin *fluere,* to flow). It is also known as fluorspar. Bluish-black to black antozonite (Pl. 20) is partly decomposed by radioactivity, with the liberation of free calcium (coloration) and free fluorine (pungent odor on fracture).

Its use in metallurgy and the manufacture of hydrofluoric acid constitute about 90 per cent of the total consumption. Clear, optical-grade fluorite is used for making lenses.

Formation and occurrence. Fluorite is found in hydrothermal veins commonly associated with lead-zinc ores and usually accompanied by quartz, barite and calcite.

ENGLAND – numerous localities for superb crystals (!!!) of fluorite in a great variety of color, the best known being blues, purples, and greens; some localities are lead-zinc mines around Weardale, Durham Co., Alston Moor, Cumberland, and Castleton, Derbyshire, known for massive purple to colorless banded fluorite formerly much used for ornaments ('blue-john'). USA – crystals from 2 cm. to 30 cm. across (!!!) mostly purple, but also yellow, blue, etc. from fluorite mines in the region around Cave-In-Rock, Illinois and adjacent portions of Kentucky; large green crude crystals, often very clear, Westmoreland, Cheshire Co., New Hampshire; fine brown crystals in limestone cavities around Clay Center, Ottawa Co., Ohio; pale green octahedral crystals (!) with pink rhodochrosite from Sunnyside mine, Silverton, Colorado; fine blue crystals in Hansonburg district, Socorro Co., New Mexico. GERMANY – crystals (!!) in Grube Luise, at Stolberg, Harz; crystals (!) in Grube Clara near Wolfach and Grube Friedrich Christian in Wildschapbachtal, Baden-Württemberg; mines near Münstertal, south of Freiburg, crystals (!); numerous occurrences of crystals (!–!!!) in mines of the Wölsendorf/Oberpfalz in Bavaria. SWITZERLAND – famed for pink octahedrons (!!!) implanted on quartz, feldspar, etc. in the region between Grimsel and Fellital of the Central Aar Massif (Göschener Tal, Alpjahorn, Zinggenstock, Juchli, etc.). ITALY – green crystals (!!) Val Sugano, Trentino. CANADA – sharp clear crystals (!–!!!) from Madoc area, Hastings Co., Ontario.

Fluorite is abundant in fine specimens, especially from the Illinois-Kentucky area, also from New Mexico, and lately from

Mexico; the pink octahedrons of Switzerland are much prized and astonishingly expensive! English and German specimens are also available, mostly through dispersal of old collections.

CRYOLITE

Na_3AlF_6, monoclinic; very rare from just a few localities. At Ivigtut, SW Greenland, however, there is an outcrop of the mineral of about $115 \times 50 \times 70$ m., but the occurrence has been completely exhausted. The deposits sit on top of a granite boss and were mined for aluminum smelting.

Class 4 Oxides and hydroxides

The oxide minerals are composed of one or more metals with oxygen; the hydroxides consist of one or more metals with hydroxyl (OH) groups.

CUPRITE

CHEMISTRY Cu_2O, copper oxide.
STREAK reddish-brown   CUBIC   H $3\frac{1}{2}$–4   SG 6
COLOR deep red, but usually dark brownish red, often with purplish cast.
LUSTER adamantine to submetallic.
PROPERTIES distinct octahedral cleavage, conchoidal fracture, brittle, transparent to translucent.
CRYSTALS octahedra and cubes and complex modified crystals but rarely in sizes over several mm.; mostly massive granular.
IDENTIFICATION color, very commonly on or with native copper and copper minerals; crystal shapes.

Cuprite comes from the Latin *cuprum*, copper. It is the richest ore of copper, but because of its almost exclusive occurrence in the decomposition or secondary enrichment zones, most deposits have already been exhausted.

Formation and occurrence. Its formation is limited to the lower regions of oxidation zones; native copper, malachite and azurite are formed simultaneously.

FRANCE – splendid octahedral crystals (!!!) to 3 cm. diameter, coated with malachite, at Chessy, near Lyon. ENGLAND – crystals (!–!!!) from sulphide veins near Redruth and Liskeard,

Cornwall. USSR – very fine sharp crystals (!!!) obtained years ago from Bogoslovsk, Ural Mts. USA – crystals (!–!!) from Bisbee, Cochise Co., Arizona, also the fibrous bright-red type known as *chalcotrichite;* abundant on copper from Chino pit, Santa Rita, Grant Co., New Mexico. Crystals over 1–2 cm. are rarely seen and command high prices; the Chessy specimens are particularly desired.

ZINCITE

ZnO, only economic occurrence at the well-known locality (formation controversial) of Mine Hill, near Franklin, and Sterling Hill, near Ogdensburg, Sussex Co., New Jersey. With franklinite, gahnite, willemite Zn_2SiO_4, smithsonite, calcite, etc.

The spinel group

Members of the spinel group all have a general formula AB_2O_4. A is a divalent metal and may be Mg, Fe^{++}, Zn or Mn; B is a trivalent metal, Al, Fe^{3+} or Cr. Various solid solution series are possible. For instance:

aluminum spinels	*iron and chrome spinels*
spinel $MgAl_2O_4$	magnetite $FeFe_2O_4$
pleonaste (ceylonite)	franklinite $(Zn, Mn, Fe)Fe_2O_4$
$(Mg, Fe)(Al, Fe)_2O_4$	
hercynite $FeAl_2O_4$	chromite $FeCr_2O_4$
gahnite $ZnAl_2O_4$	picotite $(Mg, Fe)(Al, Fe, Cr)_2O_4$

All spinels crystallize in the cubic system, mostly in octahedra, less frequently in rhombic dodecahedra. Twinning according to the spinel law is common (Fig. 7). With the exception of chromite all spinels have indistinct cleavage, and fracture is always conchoidal. Aluminum spinels are transparent, in thin flakes at least, whereas thin flakes of iron and chrome spinels are only translucent.

SPINEL

CHEMISTRY $MgAl_2O_4$, magnesium aluminum oxide; iron, chromium, etc. may substitute in the structure.

STREAK colorless   CUBIC   H 8   SG 3.7

COLOR red, pink, blue, purple, green, brown, etc.; very rarely colorless.

LUSTER glassy.

PROPERTIES conchoidal fracture, sometimes uneven; brittle but tough; transparent to opaque.

CRYSTALS usually well-developed octahedrons with edges modified by narrow faces; also twins ('spinel law'); rounded grains.

IDENTIFICATION crystal shape, hardness, color and transparency.

The origin of the name is not known, although it was used by Agricola. Distinct varieties: gem-quality spinel is red or slightly pink owing to a little Cr ('balas ruby' or 'spinel ruby'); blue spinel with a little Fe^{++}; green spinel (chlorospinel) with much Fe^{3+} and a trace of Cu. Good stones of over 4 carats are rare. The largest known cut spinels weigh 361 and 400 carats. Spinels are also produced synthetically.

Formation and occurrence. Formed almost exclusively as a result of contact-metamorphism in limestones or dolomites, rarely in granites and pegmatites, also in gneisses. Because of its hardness it is a common constituent of gravels.

CEYLON – splendid small and sharp crystals much used for gems; smaller crystals (!!!) used for micromounts. BURMA – similar occurrence to Ceylon. CANADA – crystals (!!) of opaque spinel in marbles, 2–4 cm. diameter, in Grenville rocks of Renfrew and Leeds counties, Ontario. NEW YORK – similar Grenville rocks yield spinels in St Lawrence Co.; exceptionally large and sharp black crystals (!!!) from marbles in Orange Co., the occurrences extending southward into NEW JERSEY at Franklin, Sussex Co., where crystals (!!!) have been found to 12 cm. on edge. MASSACHUSETTS – good crystals and aggregates at Rowe, Franklin Co. SWEDEN – Aker, Södermanland, crystals to 3 cm. ITALY – small, very sharp crystals (!!) in limestone ejecta at Monte Somma, Vesuvius. MADAGASCAR – lately in curious ball-like crystals (!!) highly modified, from Antanimora and Ambatomainty, associated with diopside, phlogopite, etc.

PLEONASTE (or Ceylonite)

Black, in ultrabasic rocks, also in metamorphic rocks and gravels.

GAHNITE

Dark green to black, occurs in garnet-cordierite gneisses in Sweden; at Bodenmais, Bavaria; in New Jersey at Franklin and Ogdensburg. Distinctive greenish or bluish.

MAGNETITE   Plate 13

CHEMISTRY Fe_3O_4, iron oxide; substitutions result in other species in the series, e.g., magnesioferrite ($MgFe_2O_4$), franklinite ($ZnFe_2O_4$), jacobsite ($MnFe_2O_4$), and trevorite ($NiFe_2O_4$), but only magnetite is abundant.

STREAK black   CUBIC   H $5\frac{1}{2}$–6   SG 5

COLOR black.

LUSTER submetallic, sometimes splendant.

PROPERTIES fracture subconchoidal to uneven; sometimes good octahedral parting; brittle; opaque.

CRYSTALS crystals uncommon as octahedrons, dodecahedrons or combinations of these forms; mostly granular massive.

IDENTIFICATION magnetism, the only other magnetic mineral being pyrrhotite; crystal form; lesser hardness than spinel; chromite has brown streak; ilmenite and hematite not magnetic.

Several ancient authors mention a stone *magnes* or *magnetis,* which attracts iron. Tales from the Arabian Nights describe magnetic mountains capable of pulling out iron nails from ships' planks. It is likely that the name is derived from Magnesia, a locality bordering on Macedonia. Magnetite, containing 72 per cent Fe, is the most valuable ore of iron.

Formation and occurrence. A common accessory mineral in basic igneous rocks; as a result of liquid magmatic segregation large deposits of titanomagnetite may be formed (with limonite, spinel, olivine) as well as pure magnetite deposits. A minor constituent of hydrothermal veins and alpine fissures. Magnetite may be formed by metamorphic processes from other iron minerals; it is also a skarn ore. In gravels and sands as a detrital mineral magnetite is often associated with gold, but may also form sands on its own.

Magnetite is very widespread and commonly occurs in large deposits but only those localities furnishing crystal specimens are mentioned here. CANADA – large crystals at Faraday,

Hastings Co., Ontario, but seldom sharp; a large crystal of over 150 kg. recorded. USA – crystals (!!) to 3 cm., sharp and fine, from iron ores of Mineville and Moriah, Essex Co., New York; also sharp lustrous crystals (!!) at Tilly Foster mine, Brewster, Putnam Co. Small sharp crystals in prochlorite schist, with pyrite, near Chester, Windsor Co., Vermont. In Utah the iron deposits of Twin Peaks in Millard Co. and deposits in Iron Co. furnish crystals (!) to 10 cm. diameter in clusters (many altered to martite hematite). Similarly, magnetite altered to martite hematite occurs in MEXICO at Cerro Mercado, Durango. SWEDEN – crystals (!!) at Nordmark, Värmland, and Gammal-kroppa. SWITZERLAND – Riffelberg near Zermatt, sharp crystals (!!!), same at Binnatal in crystals (!!!) more than 2 cm. across. AUSTRIA – numerous localities in the Hohe Tauern and Zillertal Alps as crystals in schist; superb lustrous crystals (!!!) also occur in vugs. ITALY – crystals (!!!) Traversella, Piedmont.

FRANKLINITE

Occurs only in quantity and crystals (!–!!!) at Franklin and Ogdensburg, New Jersey, USA in the famous deposits in the Franklin marble; crystals octahedral, some to as much as 18 cm. on edge! Completely opaque and distinguished from spinel which is usually translucent on thin edges; non-magnetic.

CHROMITE

CHEMISTRY $(Mg, Fe)Cr_2O_4$, magnesium-iron chromium oxide.
STREAK brown   CUBIC   H $5\frac{1}{2}$ SG 4–4.8
COLOR black, sometimes brownish.
LUSTER dull submetallic.
PROPERTIES brittle; uneven fracture; opaque.
CRYSTALS very rare, like magnetite; usually granular massive.
IDENTIFICATION brown streak distinguishes from magnetite and franklinite; non-magnetic.

Chromium was discovered in 1797 in crocoite. At the same time chromite (chrome iron ore, chrome iron stone) was first noticed in serpentine in the northern Urals. The name is derived from the Greek *chroma*, color, since chromium compounds are often strongly colored and are used as pigments. The development of chromium steels began in 1880, but they did not really come into their own before the first world war. Today half of the chromium produced goes into the metal industry, about a third (as chromite) is used for making refractory bricks, and the rest goes to the chemical industry.

Formation and occurrence. Chromite is formed by liquid-magmatic segregation associated with basic, subsequently serpentinized, plutonic rocks. Serpentine usually accompanies chromite. All occurrences are similar to each other. USSR – Kempirsay Massif, S. Urals; Sarany, central Urals, is said to be the world's largest deposit. TURKEY – Guleman. YUGOSLAVIA – Nada and Orasa. AUSTRIA – Kraubath, Styria. RHODESIA – Great Dyke, also in gravels here. CUBA – Cromita, Oriente.

PICOTITE

Chromium and iron bearing spinel, alters to chromite. Constituent of olivine bombs in the Eifel, Germany.

CHRYSOBERYL   Plate 12

CHEMISTRY $BeAl_2O_4$, beryllium aluminum oxide.
STREAK colorless   ORTHORHOMBIC   H $8\frac{1}{2}$ SG 3.7
COLOR greenish-yellow, yellow, olive to emerald green.
LUSTER vitreous.
PROPERTIES distinct prismatic cleavage; conchoidal fracture; brittle but tough; transparent to translucent.
CRYSTALS tabular to prismatic, common; also commonly twinned, sometimes forming sixling twins of hexagonal outline; also massive, granular.
IDENTIFICATION hardness, harder than beryl which it resembles; shape of crystals and twins.

The name comes from the Greek *chrysos,* gold i.e., a gold-colored beryl. Alexandrite and cymophane are very highly valued gem varieties. Alexandrite named after the Russian Tsar Alexander II, is colored green by Cr_2O_3, but it appears columbine-red by artificial light. Large alexandrites may fetch very high prices. Cymophane (from Greek *kyma*, a wave) is yellow-green or brown, translucent. With clever cutting in the appropriate direction a light-colored line appears which glides over the cabochon-cut stone when rotated. Such stones are called cat's-eyes. Good cymophanes have been known to fetch up to $1,000 a carat from connoisseurs.

Formation and occurrence. Mainly in granite pegmatites, also in mica schists and in contact zones between granites and schists, and in gem gravels. USSR – alexandrite sixling twin crystals (!!!) in the emerald mines of the Takovaya River district, Ural Mts.; some crystals to 8 cm. diameter; small cut gems showing strong color change were povided from clear areas. CEYLON and BURMA – waterworn gem crystals in gravels; Ceylon gravels fine gem quality cat's eyes (!!!), alexandrites (!!!) and clear faceted gems. BRAZIL – gem gravel occurrences, sometimes in crystals (!!), also gem grade material, from Minas Gerais and Espirito Santo. RHODESIA – dark twin crystals (!!) from Novello Claims, with small cuttable areas. MADAGASCAR – flat, greenish-yellow sixling twins (!!!) from Lake Alaotra; in pegmatites as at Miakanjovato. CZECHOSLOVAKIA – crystals in pegmatites, Marschendorf. USA – crystals in pegmatites in Oxford Co., Maine and at Greenfield, Saratoga Co., New York; in large rude crystals from pegmatite near Golden, Jefferson Co., Colorado.

The sixling twins from the Urals localities are especially prized for their forms, as are the Rhodesian specimens.

CORUNDUM   Plate 12

CHEMISTRY Al_2O_3, aluminum oxide; commonly contains some iron; for other impurities see varieties below.

STREAK colorless   HEXAGONAL (trigonal)   H 9   SG 4

COLOR colorless, white, blue, red, yellow, green, purple, grey and many intermediate shades and tints.

LUSTER vitreous; silky in chatoyant varieties; also pearly.

PROPERTIES fracture conchoidal to uneven; often displays very flat parting which resembles cleavage; brittle but tough; transparent to translucent; sometimes opaque.

CRYSTALS ordinary common varieties with barrel-shaped crystals of hexagonal cross-section; colored varieties commonly tapered at both ends; ruby mostly tabular crystals with hexagonal outline; also granular, massive.

IDENTIFICATION hardest mineral below diamond; shape of crystals; colors; partings; crystals often encrusted with mica-like platelets.

The word 'corundum' comes from Sanskrit or Hindustani; the origin of 'sapphire' is not known, while 'ruby' comes from the

Latin *rubens,* red. For ruby in ancient times, see garnet, p. 195. The *sappheiros* of Theophrastos is lapis lazuli (*cf.* p. 172), although some genuine sapphire must have found its way into Mediterranean lands amongst other gemstones from India. Red and blue *jacut* (an Arabic word), mentioned by Aristotle, was almost certainly sapphire, and was supposed to protect its wearer from plague and sorcery.

Common corundum is dull, usually grey or bluish-grey. The surface of the often unattractive-looking, sometimes quite large crystals is frequently altered into mica.

Gem corundum : sapphire and ruby are the best-known varieties and occur either in clear crystals or containing very fine tube-like inclusions or 'silk' which impart either cat's-eye effects or star effects in properly cut cabochon gems. Sapphire owes its blue color to small quantities of Fe or Ti; ruby to Cr. Gem corundum in other colors, orange (known as padparadschah and highly valued), green, pink, purplish to rich purple, pale to deep yellow, etc., are all called 'sapphire', preceded by the appropriate color term. While large clear blue sapphires are known (in excess of 200 carats), clear or even reasonably clear natural rubies are extremely scarce and seldom exceed several carats in weight. Synthetic sapphire and ruby, as well as other colored varieties, are manufactured in large quantities for jewelry gems, watch jewels, and technical and scientific purposes.

Emery: usually included as a variety, is not really a mineral at all, but a rock of variable composition containing corundum, magnetite, some hematite, and quartz, all in a tough fine-grained aggregate; it has been much used for abrasive applications.

Formation and occurrence. Corundum of magmatic origin is found in syenites and some pegmatites, but is also found in zones of enrichment as a result of differentiation; also present in some basaltic rocks, and in metamorphosed sediments as gneiss and marble, the last rock providing the finest crystals and gem kinds. Highly resistant to wear, it is common in gem gravels as grains and crystals. Emery is always of metamorphic origin, usually derived from aluminum-rich constituents.

BURMA – ruby crystals, usually crudely formed, in gem gravels of Mogok district; also crystals of sapphire (!–!!).

THAILAND – crystals (some interesting 'rosettes' !!) in Chanta-
bun and Krat provinces. KASHMIR – crude blue and white
crystals found in pegmatite, now exhausted, from which very
fine gems were cut. CEYLON – rude to sharp crystals (!–!!!) from
gem gravels of Ratnapura and Rakwana. INDIA – crude crystals
of plum-red, only translucent, cuttable to poor-quality star
stones. AUSTRALIA – numerous gem gravel deposits yielding
dark blue, yellow and green stones, in Queensland and New
South Wales. USA – very small gem crystals from Yogo Gulch,
Judith Basin Co., Montana, and stubby hexagonal prismatic
crystals from Rock Creek, Granite Co. in same state. Large
masses and rude crystals from North Carolina, South Carolina
and Georgia in metamorphic rocks of the Appalachian
Mountains. CANADA – crystals (!–!!) in syenites of Bancroft,
Hastings Co. and Craigmont, Renfrew Co., Ontario, some in
excess of 25 kg.! TANZANIA – Longido, northwest of Kiliman-
jaro, attractive ruby crystals in massive green zoisite (see Pl. 12).
SOUTH AFRICA – large crystals in Cape Province and Transvaal.
MADAGASCAR – large crystals.

Ruby crystals from Burma are highly prized even when un-
cuttable; small but sharp and fine crystals commonly obtainable
from lots of Ceylon gem gravel; common corundum crystals
are easily obtainable.

HEMATITE   Plate 13

CHEMISTRY Fe_2O_3, iron oxide; some titanium and magnesium
   isomorphously substitute for iron; two modifications exist:
   hexagonal (trigonal) hematite, and the much less common
   cubic maghemite.
STREAK red to blackish-red; distinctive.   HEXAGONAL (trigonal)   H $6\frac{1}{2}$ SG 5.1
COLOR steel-grey to iron-black; in finely-divided earthy masses
   a dull purplish-red.
LUSTER metallic to dull and earthy.

PLATE 23   *Calcite*

Above Calcite crystals consisting of low and very steep rhombohedrons
(*cp*. Fig, 20, p. 161). From the dolomite at Eichstätt, Middle Franconia,
Germany. *Center* Crystals of calcite which originally grew as normal
scalenohedrons, and later changed their form completely. Lenggries, Bavarian
Alps, Germany. *Below left* Parallel intergrowths of calcite. Rohrdorf, Bavaria,
Germany. *Below right* Fibrous calcite. Harburg, Bavaria, Germany.

PROPERTIES fracture conchoidal to uneven; excellent parting noted in some varieties; compact fibrous varieties splintery as in English 'kidney ore'; brittle but tough; opaque except in very thin splinters or platy crystal inclusions in other minerals.

CRYSTALS compared to the huge quantities of ordinary hematite, crystals are rare but are commonly obtainable as tabular, sharp-edged individuals with many small shining faces, also in squat rhombohedral crystals, thin tabular plates and plates in the form of rosettes ('iron roses' or 'eisenrosen'), and rarely as cube-like or simple hexagonal prismatic crystals; aggregates of many minute flaky crystals are known as 'specularite', 'specular iron ore', etc. Radiating fibrous kinds with distinctive rounded surfaces are called 'kidney ore' from their shape. Also earthy massive, and pisolitic.

IDENTIFICATION the distinctive red streak easily identifies hematite among similar-appearing magnetite, ilmenite, chromite, etc.; not magnetic; hard; crystals distinctive; titanium-rich hematites may have a black streak.

Theophrastos likened hematite (which he called *haimatites*, i.e. bloodstone) to congealed blood.

Hematite is the most abundant iron ore and has been in use for a very long time. Good kidney ore is cut and polished, and is very suitable for engraved gems.

Magmatic formation, veins. Hematite is a minor constituent of many igneous rocks, sometimes in visible flakes. Finely divided hematite is an important coloring agent in many minerals, pink feldspars for instance. It is common in hydrothermal veins with quartz, barite, magnetite, siderite, chlorite etc. The famous 'iron roses' are formed in alpine fissures. Hematite is not infrequently found as a sublimation product.

Sedimentary formation. Most mineable hematite ore is taken from sediments, the oolites for example, which, further, are

PLATE 24   *Strontianite, aragonite, siderite and dolomite*

Above Sheaves of strontianite crystals. Bruch on the Mur, Austria. *Center* Arborescent aragonite on limonite, 'flos ferri'. Eisenerz, Austria. *Below left* Crystals of dolomite, in a combination of rhombohedron and basal pinacoid. Trieben, Austria. *Below right* Crystals of siderite with curved edges and faces, and 'parquet' structure. Hüttenberg, Austria.

frequently metamorphosed. Occasionally the iron may be derived from subterranean volcanic eruptions. Large masses of sediments (sandstone, red clays) are colored red by hematite.

Metamorphic formation. Hematite is often a minor constituent of crystalline slates, or may even become a chief constituent, as in the quartz-rich rocks known as itabirites, and becoming taconites with increasing magnetic content. Hematite is also a skarn mineral.

ENGLAND – Cleator Moor, Cumberland, sharp crystals with quartz; also from other mines in Cumberland and elsewhere in England, 'kidney ore' (!!!) in masses to 25 kg. SWITZERLAND – splendid iron roses (!!–!!!) from less than a centimeter to several cm. diameter in alpine vugs at Reckingen, Aar Massif and other points, also in the Gotthard Massif as at Piz Lucendro, La Fibbia, Monte Prosa, Cavradischlucht (!!!), and in the Binnatal at Binneltini and Ritter Pass (!!!). GERMANY – crystals in Neue Hardt quarry, Siegerland. AUSTRIA – crystals and roses in the Hohe Tauern. FRANCE – crystals (!!) at Puy-de-Dôme and Puy-de-la-Tache, Auvergne. ITALY – very fine groups of complex crystals (!–!!!) in the iron ore deposits of Elba at Rio Marina. ROMANIA – crystals (!!) Dognaska, Banat. BRAZIL – large 'iron roses' and crystals (!–!!!), often of large size (to 10 cm. diameter) at Itabira, Minas Gerais. MEXICO – hematite replacements of octahedral magnetite at Cerro Mercado, Durango (!). USA – hematite (martite) replacements of magnetite at Twin Peaks, Millard Co. (!!); sharp tabular brilliant crystals near Bouse, Arizona; druses with quartz from mines near Edwards, Gouverneur and Antwerp, St Lawrence Co., N.Y.; hematite pseudos after siderite in pegmatite cavities of Crystal Peak area, Teller Co., Colorado.

ILMENITE

CHEMISTRY $FeTiO_3$, iron-titanium oxide, always a little Fe_2O_3, often interstratified with exsolution lamellae of magnetite as well as rutile.

STREAK black HEXAGONAL (trigonal) H 5–6 SG 4.7
COLOR iron-black.

LUSTER submetallic, dull.

PROPERTIES no cleavage; conchoidal to uneven fracture; parting on basal plane; brittle; opaque.

CRYSTALS commonly thin to thick tabular with hexagonal outline and edges modified with rhombohedral faces; also ball-like with rhombohedral faces; mostly massive, granular.

IDENTIFICATION resembles magnetite, hematite, etc. but lacks magnetism and streak distinguishes from hematite; powder sometimes weakly magnetic.

In 1791 a new metal was discovered in ilmenite (menaccanite) which Klaproth named titanium. Ilmenite has many local variants: ilmenite from the Ilmen Hills, Urals; menaccanite from Menaccan, Cornwall; iserine from the Iser river, Bohemia; kibdelophane from Gastein, etc. For nigrine, see under rutile, p.138.

Titanium and titanium oxide are now being produced, in addition to iron, from previously unsuitable ilmenite, titano-magnetite, and hematite-ilmenite ores, as well as from rutile. Titanium has recently become an important constituent of alloys used for extremely heat-resistant alloys. Titanium dioxide (titanium white) has very high covering power, and today completely replaces the white paints (lithopones) hitherto made from sphalerite and barite.

Formation and occurrence. Ilmenite is a common minor constituent of basic igneous rocks. As a result of liquid magmatic segregation large ore bodies of ilmenite may be formed with or without titanohematite, or together with hematite. Large crystals of ilmenite occur in certain pegmatites, and it is frequently found in alpine fissures, partly as 'iron rosettes'. Titanium sands are widespread; associated minerals are magnetite and hematite, garnet, apatite, rutile, spinel, plagioclase, augite, all minerals of alpine fissures.

USSR – large crystals (!!) near Miass, in Ilmen Mountains, Urals. SWITZERLAND – beautiful rosettes in Maderanertal. FRANCE – crystals at St Christophe, Bourg d'Oisans. NORWAY – large rhombohedral crystals (!!!) at Kragerö, Froland to 13 cm. diameter. USA – crystals in marble and serpentine in Orange Co., New York.

Despite abundance in massive forms, crystals are much rarer than would be supposed; those from Norway are largest.

The quartz group   Plates 14–18, also 6, 19, 34, 47

With the exception of opal, which apparently has a variable water content (SiO_2 + aq.), the chemical composition of quartz mineral is SiO_2.

The whole group is listed in Table 9 for completeness; only quartz and opal are common and therefore described in detail.

TABLE 9 The Quartz Group

	crystal system	SG	H
quartz	trigonal	2.65	7
beta-quartz	hexagonal	2.53	?
tridymite	orthorhombic	2.27	$6\frac{1}{2}$–7
high-tridymite	hexagonal	2.20 approx.	?
cristobalite	tetragonal	2.20	$6\frac{1}{2}$
high-cristobalite	cubic	2.20 approx.	?
coesite	monoclinic	3.01	$7\frac{1}{2}$
stishovite	tetragonal	4.28	?
lechatelierite	amorphous	2.20	7
keatite	tetragonal	2.50	?
opal	amorphous		$5\frac{1}{2}$–$6\frac{1}{2}$

Beta-quartz is stable only above 573°C; crystals formed above this temperature retain their characteristic form on cooling: hexagonal bipyramid and very short (or absent) prism. *Tridymite* and *cristobalite* may occur occasionally as tiny crystals in volcanic rocks such as andesites or trachytes, as at San Cristóbal near Pachuca, Mexico, or in the Euganean Hills of north Italy (paramorphs of quartz after tridymite), or in the Siebengebirge, Germany. A few years ago *coesite, stishovite* and *keatite* were prepared synthetically at very high pressures; minute crystals of these minerals were subsequently found naturally in meteor craters (Meteor Crater, Arizona; Nördling Ries, Bavaria). *Lechatelierite* is natural quartz glass which is sometimes formed

PLATE 25   *Malachite and azurite*
Above Radiating, botryoidal malachite on fluorite. Lichtenberg, Upper Franconia, Germany. *Below* Elongated, platy crystals of azurite with malachite. Tsumeb, Otavi district, South-West Africa.

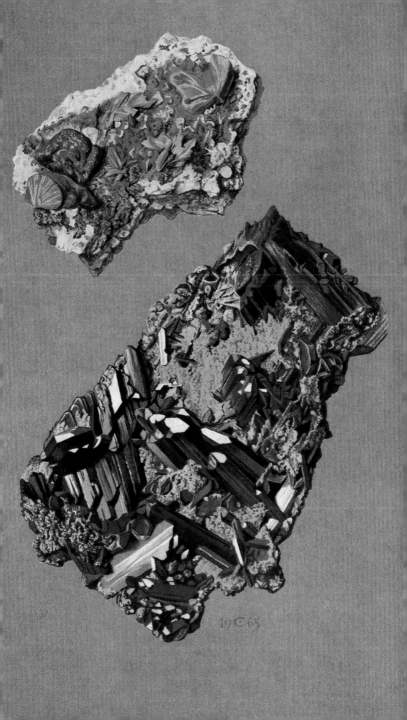

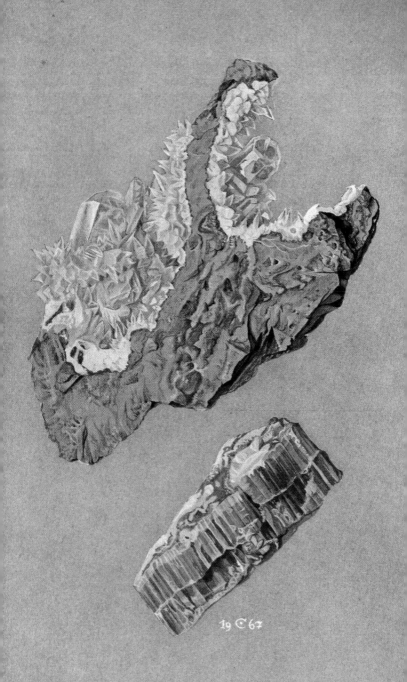

19 C 67

as a result of lightning striking quartz sands, and as wind-worn pebbles of unknown origin in the Libyan desert.

QUARTZ   Plates 14–18 and 6, 19, 34, 47

CHEMISTRY colorless clear crystals are predominantly SiO_2, only traces of Al, Na etc. substitute; considerable impurities by inclusion of gases, liquids and other minerals (Table 10).

STREAK colorless   HEXAGONAL (trigonal)   H7   SG 2.65

COLOR colorless, purple, brown to nearly black, rose-colored, yellow, blue; massive forms and compact varieties white, or any other color, often very strongly colored.

LUSTER vitreous, resinous on fractures.

PROPERTIES poor cleavage parallel to rhombohedral planes, fracture conchoidal, brittle (in crystals), transparent to translucent.

CRYSTALS extremely common (see Figs. 17–19). Hexagonal prism, and two rhombohedra which together give the impression of a pyramid, are the principal forms. Additional forms are: steeper rhombohedra, trapezohedral faces, pyramids. Right-handed and left-handed quartz crystals may be distinguished from each other by the positions of the trapezohedral faces (if present). Two kinds of twinning are common: in *Dauphiné twins* either two left-handed or two right-handed crystals are combined, and in *Brazil twins* a left-handed crystal is combined with a right-handed one. The twins are fully interpenetrant so that no re-entrant angles appear. *Japanese twins* are two individuals joined at nearly right angles. Massive, granular to compact (cryptocrystalline), radiating, fibrous.

IDENTIFICATION crystal form, hardness, absence of cleavage when massive (distinction from feldspar).

Quartz is the most abundant mineral, after the feldspars. It exhibits a greater range of colors and varieties than any other mineral. The crystalline varieties and the cryptocrystalline compact chalcedonies form two separate groups. For clarity

PLATE 26   *Celestite*
Above Tabular celestite crystals with calcite scalenohedrons in limestone. La Reuchenette near Biel, Swiss Jura. *Below* Columnar celestite as a fissure filling. Dornburg near Jena, E. Germany.

TABLE 10 Varieties of quartz, with colors and inclusions

Crystalline and massive varieties

Name	Habit	Color	Transparency	Cause of coloring, inclusions etc.
quartz, common vein quartz	massive, crystals	white, grey yellowish	dull, opaque	gases and liquids, cracks etc.
rock crystal	crystals	colorless	transparent	—
smoky quartz	crystals	brown	transparent	coloring caused by radiation
morion	crystals	brown to nearly black	translucent	coloring caused by radiation
amethyst	crystals, also massive	purple	transparent	coloring caused by radiation (? Fe)
citrine	crystals	yellow	transparent	traces of very finely divided FeOOH
rose quartz	massive, crystals rare	pink	translucent	minute rutile needles (? Mn)
blue quartz	granular, massive	blue	transparent to translucent	minute rutile needles
prase	massive	leek-green	translucent	abundant actinolite needles
aventurine	massive	iridescent, various colors	opaque	flakes of mica or hematite
tiger's eye	fibrous	blue to golden	opaque	silicified crocidolite asbestos

these are summarized in Table 10. A few particular characteristic crystal forms have been given separate names: scepter-
quartz and skeletal-quartz (Pl. 16), Babel-quartz, where the
prisms form steps, or the 'gwindels' (Pl. 16), nearly parallel,
slightly twisted twinned intergrowths, are examples of such
names. In phantom quartzes (Pl. 16) the crystals were dusted
over, maybe more than once, by a layer of chlorite or clay, and
then continued to grow again. This 'dust layer' imparts a
ghost-like appearance to the interior of the crystals. Many

Compact varieties

Name	Habit	Color	Transparency	Cause of coloring, inclusions etc.
chalcedony	compact	pale blue, pale grey	translucent	evenly colored or slightly banded
cornelian	compact	yellowish to deep red	translucent	very finely divided hematite
sardonyx	compact	brown	translucent	very finely divided iron hydroxide
chrysoprase	compact	apple-green	translucent	hydrated nickel silicates
agate	compact	multi-colored	translucent	finely banded, filling cavities
onyx	compact	grey, black and white	opaque	bands thicker than in agate
enhydros	compact	as chalcedony	translucent	cavities partly filled with water
jasper	compact	multi-colored	opaque	many impurities of clay, Fe etc.
plasma	compact	leek-green	opaque	chlorite, hornblende and other green minerals
heliotrope	compact	green with red flecks	opaque	red flecks due to hematite
flint	compact	white, grey and other colors	opaque	like jasper, containing some opaline material

crystals from alpine fissures have a thick powdery coating of chlorite. In 'capped quartz' these layers, separated by impurities, may sometimes become clearly discernible. Clear rock crystal may often contain bundles of rutile needles as inclusions (Pl. 19).

Quartz was well known by the ancients, and the six-sided prisms of rock crystal were described in detail by Pliny. It was generally believed that rock crystal was very deep-frozen ice, which could no longer be re-melted. In order to prove this,

Pliny recorded that 'krystallos' was found in the vicinity of glaciers. He also reported that, in Rome, luxurious vessels were carved from it. Agate, sardonyx, chalcedony, prase, onyx, heliotrope and other names can all be found in ancient writings, but the precise meaning attached to them is not always clear. Carnelian was one of the most popular gem-stones in those days, but all varieties were very widely used, as they still are today. During the sixteenth and seventeenth centuries, large, heavily ornamented decorative vases were cut from rock crystal. For a long time rock crystal used to be brought by the barrel from Brazil and Madagascar to Idar-Oberstein in the Rhineland, the great gem-cutting center. Good amethysts are in demand for rings, and fetch quite high prices. Rose-quartz used to be very popular, and was made into necklaces. Plasma, onyx with its black and white bands, and agates are very suitable for artistic cutting. The prized black color, however, is obtained by artificial coloring, and in any case, artificial dyeing and the misnaming of some chalcedonies and of quartzes has gone on for a very long time. Thus 'oriental topaz' is merely heated low-grade amethyst, and 'Scotch topaz' is ordinary smoky quartz (cairngorm); these names are given to enhance their commercial value.

Quartz has many industrial uses: untwinned rock crystal is used in the electrical and radio industries; agates for making mortars and bearings; glass and porcelain are made from very pure quartz sand, as also are quartz glass for ultra-violet lamps, and various silicones.

Formation and occurrence. Quartz is extremely abundant in veins and also as a rock-forming mineral (see Tables 5–7); it is the commonest vein mineral associated with ores. Well-developed crystals occur mainly in alpine fissures and other types of veins, occasionally also in other rocks. Quartz also occurs as a clastic mineral or as hardstone (hornfels), and in sediments. Agate, together with amethyst and zeolites, is one of the typical minerals filling amygdales in basic lavas; chalcedony may often be formed by the decomposition of other rocks, and not infrequently as a dehydration product of opal. All types of quartz may be found in gravels, shingles, sands, etc.

SWITZERLAND – Rock crystal (!!!) and smoky quartz (!!!) in many alpine fissures, often without any other associated

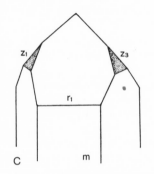
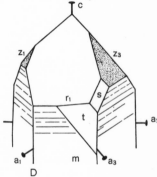

17 Quartz. (A) A simple crystal of quartz appears to consist of a hexagonal prism m and of a hexagonal pyramid; the pyramid, however, is the combination of two juxtarotated rhombohedra r and z. The faces r_1 to r_3 are completed by the broken lines to the full rhombohedron r. (B) Rhombohedron z. (C) Trigonal crystal of quartz. The rhombohedral faces z are much less well developed than those at r and are very rough. (D) Typical combination present in rock crystal (Pl. 14, top). In addition to the faces m, r and z, the trapezohedron t and the trigonal pyramid s are also present. The crystal illustrated is right-handed quartz since t and s lie on the right. In a left-handed crystal these faces would occur on the left. The striations on m are due to steeper rhombohedra (*cp*. Fig. 19).

minerals. Occurs in two habits: normal crystals with straight prisms; 'Ticino' type (Fig. 19), steeply wedge-shaped. Its chief occurrence is in the Aar Massif (W. of Aar to E. of the Reuss), particularly at Zinggenstock, the Rhône and Tiefen glaciers, Fellital, Val Giuv, Maderanertal, etc. In 1719 a cavity was found at Zinggenstock containing over 50 tons of crystals, and a morion crystal 69 cm. long, weighing 133 kg., was found at Tiefen in 1868. At Val Tremola, Ticino, are large Ticino-type

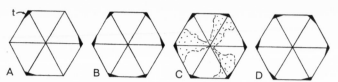

18 Quartz twins in cross-section. (A) Untwinned left-handed quartz crystal. All rhombohedral faces are drawn to equal size. The small black triangles indicate the trapezohedra. (B) Right- and left-handed crystals are intergrown as very thin lamellae, not illustrated (Brazil twin). (C) Two left-handed crystals irregularly intergrown (Dauphiné twin). (D) As C, but with two right-handed crystals.

crystals, as also at the Greina Pass, and Piz Aul, Graubünden, which is a classical locality for rock crystal with rutile needles. GERMANY – 'Rhinekiesel', water-worn rock-crystal pebbles, in the Rhine; at Suttrop and Warstein near Lippstadt, North Rhine-Westphalia, loose crystals (!) (as Fig. 17); at Idar-Oberstein, Rhineland-Palatinate, agate nodules (!!) with amethyst, common quartz, zeolites, calcite, etc. Agate polishing has been carried out at Idar since 1500, and gem cutting for nearly as long; emigrants from Idar discovered the agate occurrences of Uruguay and Rio Grande do Sul in 1827. At Usingen in the Taunus, Hesse, large radially intergrown crystals, often capped; 'Pfahl' in the Bavarian Forest, where an enormous quartz vein, over 100 yards wide, runs through the whole forest from NW. to SE.; Hagendorf, Bavaria, quartz crystals up to 10 cm. long in a feldspar quarry; Epprechtstein (*cp.* feldspar, p. 247); Göpfersgrün, in the Bavarian Fichtelgebirge, stellate quartzes and pseudomorphs of steatite after quartz; Frankenstein and Jordansmühl, Silesia, chrysoprase. AUSTRIA – alpine fissures in the Hohe Tauern, Carinthia and Salzburg, rock crystal (!!!) and smoky quartz; near Mallnitz, in the moraines of the Pasterzen glacier, where in the last century a doubly terminated crystal was found weighing 25 kg.; Rauristal, rock crystal and smoky quartz (in 1811 a crystal of nearly 100 kg.); Habachtal, smoky quartz crystals up to 60 cm.; Hüttenberg, Carinthia, chalcedony; Weitendorf, Styria, aventurine boulders; Grubach, near Kuchl, Salzburg, blue quartz with riebeckite (crocidolite); Zillertal Alps, Tirol, in alpine fissures, amethyst.

ITALY – Tiso and Valle di Fassa (see zeolites, p. 253), 'Theis nodules' (agate amygdales with rock crystal and amethyst);

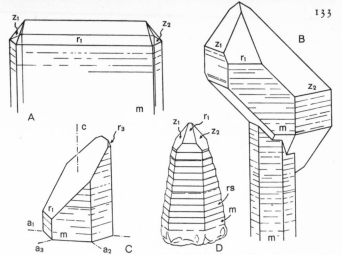

19 Distorted quartz crystals and abnormal forms (crystal faces are labelled as in Fig. 17). (A) Tabularly distorted crystal with apparently orthorhombic symmetry. (B) Scepter-quartz, somewhat like Pl. 16, top right. The crystal sitting on top is diagonally distorted. (C) Dauphiné habit rock crystal. One of the rhombohedral faces has grown very large at the expense of the other five, simulating monoclinic symmetry. (D) Ticino-type rock crystal: by the alternate development of a prism and a steep rhombohedron a tapering crystal is formed, characteristic of localities in Ticino. This alternation of prism and rhombohedron usually presents itself as striations in most quartz crystals.

Monte Tondo, Vicenza, Veneto, enhydros; Carrara, Tuscany, completely clear rock crystal (!!) in marble. FRANCE – Bourg d'Oisans, Dauphiné, rock crystal (!!!) with a predominant rhombohedron face. CZECHOSLOVAKIA, ROMANIA – Amethyst in gold-silver veins at Banská Stiavnica, Baia Mare, Kapnica, Baia Sprie. GREAT BRITAIN – Cairngorms, Scotland, pegmatite smoky quartz, as ornamental stone. USSR – rock crystal, smoky quartz and amethyst in aquamarine-bearing pegmatites; Sverdlovsk, Urals, aventurine (!!) in mical slates, from which at one time vases were made.

SOUTH AFRICA – Griquatown in Doorn Mountains, tiger's eye (!!!), the world's best, in golden, brown, blue and green hues, and much used in lapidary work; replaces asbestos and retains fibrous chatoyant character.

USA – large clear, cloudy, milky, and smoky quartz crystals, and rarely amethyst, from many pegmatites in the New England

States; fine rock crystal (!!!) in quartz veins in the Ouachita Mountains, Arkansas, sometimes in crystals to 30 cm. and over; rock crystal-lined geodes (!) in sedimentary rocks of Mississippi River Valley around junction of Illinois, Iowa, and Missouri; petrified palm wood along a coastal plain of southern Texas; beautifully colored petrified wood (!!!) in the Petrified Forest National Park, Arizona and adjacent areas; large amethyst crystals (!!) at Four Peaks, Mazatzal Mountains, Arizona, and Japanese twins to 25 cm. across at Patagonia, Santa Cruz Co.; also chrysocolla chalcedony (!!!) in copper mines of this state; smoky quartz crystals (!!) in pegmatites of the Pikes Peak granite, west of Denver and Colorado Springs, Colorado and fine white quartz crystals (!!) at Ouray.

MEXICO – amethyst crystals (!!!) in Guerrero; banded agates (!!!) at many points in Chihauhua; druses of amethyst (!!) in silver mines of Guanajuato.

BRAZIL – enormous quantities of agate, amethyst, enhydros, etc. in lavas of Rio Grande do Sul (extending southward into Uruguay); at Serro do Mar an agate nodule 10 m. long was found; rose quartz (!!!), rock crystal, smoky quartz (!!), and many varieties of quartz in pegmatites and vein deposits in Minas Gerais, Bahia, etc.; superb rutilated quartz (!!!) from Itabira, Minas Gerais; at Diamantina a rock crystal of $5\frac{1}{2}$ tons was found. URUGUAY – the agate region extends southward from Rio Grande do Sul in Brazil into the Salto district where enormous quantities of agate nodules and excellent amethyst crystals occur; nearly all commercial agate comes from Uruguay and Brazil.

JAPAN – Japanese twins (!!!), the classic sources being in quartz veins at Otomezaka, Yamanashi Prefecture and Narushima Island, Nagasaki Prefecture. AUSTRALIA – superb chrysoprase (!!!), the world's best, Marlborough, Queensland.

OPAL   Plate 18

CHEMISTRY SiO_2, silicon dioxide, always containing from 1 to 21 per cent water.

PLATE 27   *Barite*
Above left 'Desert rose'. Norman, Oklahoma, USA. *Above right* Tabular crystals of barite with very well-developed 'parquet' structure. Cumberland, England. *Below* Common foliated barite in the 'cock's comb' form. Wölsendorf, Bavaria, Germany.

19 C 66

STREAK colorless   AMORPHOUS   H $5\frac{1}{2}$–$6\frac{1}{2}$ SG 1.9–2.5

COLOR colorless, milky, white, pale tan, green, blue, etc., depending on impurities; precious opal displays many hues also, predominantly greens, blues, red, and purples.

LUSTER vitreous to waxy; also dull, earthy.

PROPERTIES excellent conchoidal fracture; very brittle; transparent to nearly opaque.

CRYSTALS amorphous, but sometimes forming pseudos after crystals of glauberite (Australia).

IDENTIFICATION resembles chalcedony but much less hard and far more brittle; precious varieties unmistakeable by flashes of color.

Opal, from the Sanskrit *upala*, 'valuable stone', has been known for a long time as a gemstone, and was particularly coveted by the Romans. It was regarded as an unlucky stone.

Multicolored precious opal is translucent with an exceptionally fine play of colors. *Fire-opal* is yellow to red, fairly transparent, with or without color-play. *Hydrophane,* from the Greek *hydor,* water, and *phanos,* shining, is a light-colored dull opal, becoming transparent under water. *Hyalite,* Greek *hyalos,* glass, is water-clear, without play of colors, usually in reniform crusts. *Common opal* is opaque, usually dull, also with waxy luster, and may be any color. *Siliceous sinters* or *geyserites* are crustlike, stalactitic or pisolitic deposits in hot springs. *Wood-opal* is opalized wood. *Kieselguhr* (diatomaceous earth), polishing slate, and 'tripoli stone' are opaline rocks. Opals are cut as cabochons and black opals command a particularly high price from connoissuers.

Formation and occurrence. The skeletons of various living organisms such as diatoms, radiolaria, and some sponges consist of opal-like material which becomes deposited in sedimentary rocks. Usually rapid decomposition follows, and most of the original opal content of flints, horn-stones and siliceous limestones is almost completely altered into chalcedony. Opal

PLATE 28   *Gypsum*
Left Crystals of gypsum with sulphur. Agrigento, Sicily. *Right* Montmartre or 'fish-tail' twin of gypsum. The top face is parallel to the best cleavage direction; the presence of air between the cleavage flakes is the cause of the Newton rings that can be seen. Montmartre, Paris, France. *Below* Perfect crystal of gypsum.

may be formed in hot springs, which can also decompose and silicify any other rocks. The weathering of siliceous rocks commonly leads to the formation of opal.

CZECHOSLOVAKIA – classic precious opal deposits at Cernevica, near Prešov and around the Limonka and Sibanka hills have been worked since Roman times. USA – opalized wood, sometimes with magnificent play of color in colorless, white, to black opal occurs in clayey sediments in the Virgin Valley, Humboldt Co., Nevada, specimens reaching as much as 15 cm. across; opalized wood logs (!!!) occur widely in Washington, Idaho, Oregon, Utah, and Nevada, a locality at Clover Creek, Lincoln Co., Idaho being notable for oak limbs. MEXICO – fire opal and precious opal (!!!) in nodules in volcanic rocks of Queretaro and Jalisco; hyalite opal (!!!) Cerro del Tepozan, San Luis Potosi. HONDURAS – precious opal (!) in various localities. AUSTRALIA – most precious opal is now obtained from numerous deposits in New South Wales (Lightning Ridge and White Cliffs), White Cliffs providing precious opal replacements of fossils (!!!) and glauberite crystals (!!!), also deposits in Queensland and South Australia.

RUTILE   Plate 9

CHEMISTRY TiO_2, titanium oxide; commonly with up to 30 per cent Fe^{3+} (var. nigrine); also contains niobium or tantalum.

STREAK yellowish-brown   TETRAGONAL   H $6-6\frac{1}{2}$ SG 4.2

COLOR mostly dark reddish-black; also dark red, brownish, yellowish.

LUSTER adamantine to metallic.

PROPERTIES distinct prismatic cleavage; conchoidal to uneven fracture; brittle; transparent to opaque.

CRYSTALS commonly in stubby prisms with striated sides capped by low pyramids; also elongated striated prisms to acicular and fibrous, the more slender crystals becoming noticeably red or even coppery, golden, etc. Commonly twinned, forming cyclic twins or networks of slender crystals in twinned orientation (sagenite); also granular massive.

IDENTIFICATION bright luster distinguishes from other dark, heavy minerals; striated and twinned crystals; deep red color easily noted in splinters or thin crystals.

Orthorhombic *brookite* and tetragonal *anatase* are two further modifications of TiO_2. Brookite occurs in tabular crystals with indistinct cleavage, anatase in steep pyramids, blue-black to brown, commonly as small crystals in alpine vugs.

Titanium dioxide was discovered in the 1790's by Klaproth; Werner named the mineral from the Latin *rutilus,* reddish. Black nigrine may either be very iron-rich rutile, or ilmenite with a considerable excess of Ti. Rutile is obtained from gravels as titanium ore.

Formation and occurrence. Rutile is a minor constituent of plutonic rocks and crystalline slates, and a characteristic member of the mineral associations of alpine fissures, e.g. quartz, adularia, albite, chlorite, calcite, siderite, muscovite, ilmenite, apatite, and rutile, anatase, brookite. Basic pegmatites are also rutile-bearing, e.g. nelsonites, with rutile, apatite, biotite, chlorite. Rutile is common in gravels and sands.

USA – large brilliant crystals (!–!!!) with kyanite, lazulite, etc. at Graves Mountain, Lincoln County, Georgia to 12 cm. in length, commonly twinned; black singles and cyclic twins (!) in Magnet Cove, Hot Springs Co., Arkansas; superb sagenite (!!!) and stubby crystals (!!!) in the emerald localities around Hiddenite, Alexander Co., North Carolina; in stubby simple crystals (!!) in pyrophyllite Champion mine, near Laws, Inyo Co., California. USSR – crystals (!!) near Miass, Ilmen Mountains, Urals. NORWAY – simple crystals (!) at Lofthus, Snarum. SWITZERLAND – numerous alpine vug occurrences, notably Maderanertal, Val Tremola, Val Canaria, Cavradischlucht (small flattened brilliant crystals in oriented overgrowth on hematite crystals (!!)), in the dolomite of Alpe Campolungo and Lengenbach Quarry, Binnatal, and in the Lukmanierschlucht and on Piz Ault near Vals where splendid crystals (!!!) have been found. AUSTRIA – crystals (!!) Ankogel region, and in the Hohe Tauern. BRAZIL – outstanding 'swallowtail' twins (!!!) at Cerro Frio, Minas Gerais, and as inclusions in quartz (!!!) at Itabira.

CASSITERITE   Plate 19
CHEMISTRY tin oxide, isomorphous Fe, Mn.
STREAK yellowish-brown to white   TETRAGONAL   H 6–7
    SG 7

COLOR brown to brownish-black, occasionally yellow, gre
very rarely colorless.

LUSTER high adamantine, in very dark crystals metallic.

PROPERTIES imperfect cleavage, fracture conchoidal, britt
transparent to opaque.

CRYSTALS common, twinned, needles ('needle-tin'), foliate
knobbly ('wood-tin'), weathered rounded grains ('pebbl
tin').

IDENTIFICATION much heavier than similar rutile; streak pa
colored; brilliant luster; crystal shapes.

The word 'tin' is derived from the Latin *stannum*, though ori
inally this referred to an alloy; the Romans regarded tin as
special variety of lead. Cassiterite comes from the Gree
kassiteros, tin, which in turn can be traced back to the Sanskr
word *kastira*. The oldest tin-bronze finds have been dated
3500 BC. It is still uncertain whether the Cassiterides (Tin Isle
of the ancients were in fact Cornwall. Most tin came fro
Spain. At first, tin was obtained solely by washing tin-gravel
which were very widespread (e.g. Cornwall, Erzgebirg
Spain); only in the twelfth century was tin mined in th
Erzgebirge and in Cornwall.

Cassiterite is the only important source of tin, which
primarily used in alloys. Apart from the bronzes numerou
tin alloys exist, e.g. solder (40 per cent Sn, 60 per cent Pb
bush metals, type metal; Wood's alloy (1.5 per cent Sn, 25 p
cent Pb, 50 per cent Bi, 12.5 per cent Cd) melts at 67.5°C. Ti
was formerly widely used for utensils; today only sma
amounts are so used, and that mainly for making tube
Aluminum foil has replaced tin foil. Large quantities of tin g
into the making of tin-plate.

Formation and occurrence. Tin deposits are always associate
with acid igneous rocks. Beside the pegmatite assemblage wit
stannite, chalcopyrite, quartz, orthoclase, muscovite, rutil
beryl and columbite, the pneumatolytic cassiterite assemblag

PLATE 29 *Wolframite and wulfeni*
Above Wolframite, with shiny cleavage surfaces, on quartz. Zinnwal
Erzgebirge, Germany. *Below* Specimens of wulfenite, tabular crystals on th
left, from Arizona, USA, and equant crystals, consisting of prism and bas
pinacoid, on the right, from Villa Ahumada, Sierra de los Lamentos, Mexic

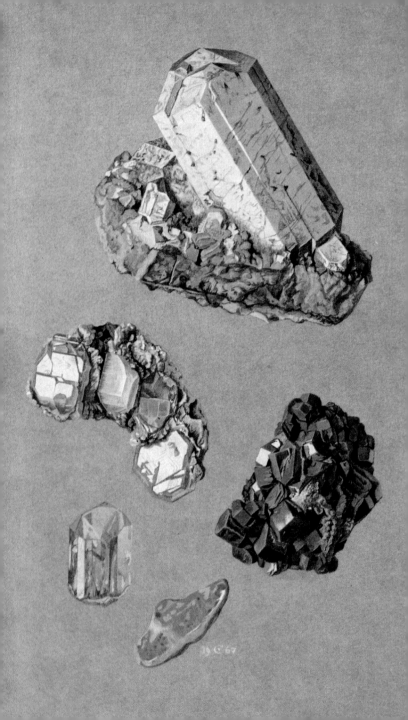

is of importance: wolframite, scheelite, molybdenite, arseno-pyrite, topaz, zinnwaldite, fluorite, tourmaline, apatite (*cp.* also p. 48). Pneumatolytic veins may grade into hydrothermal ones containing pyrite and chalcopyrite in addition to the cassiterite assemblage. Disseminated tin ores are also known. Wood-tin with hematite is a sublimation product (Mexico). Cassiterite gravels are very important economically.

BOLIVIA -- the world's largest and finest crystals (!!!) occur in the tin deposits of Llallagua, Araca, and Oruro districts. Commonly the crystals reach diameters of 2–8 cm. with un-common sharpness, smoothness of faces, and transparency; some clear zones afford very brilliant faceted gems. ENGLAND – numerous deposits in Cornwall provided specimens (!–!!) of crystals on matrix. GERMANY – fine crystals and groups (!–!!) from tin mines of the Erzgebirge at Graupen, Altenberg, Zinnwald, and Ehrenfriedersdorf in Saxony; some Schlaggen-wald crystals reached 7 cm. diameter. FRANCE – small but fine crystals (!!) at La Villeder, Morbihan. PORTUGAL – fine crystals (!!) with apatite, wolframite, siderite, etc. from Panasqueiras near Fundão. MALAYA and INDONESIA – gravel deposits, some-times with loose crystals (!). MEXICO – wood tin (!!) at many points in San Luis Potosi and Guanajuato. AUSTRALIA – druses (!!) from Emmaville, New South Wales.

It is now difficult to obtain fine crystallized specimens except from old collections; consequently the prices asked are very high, especially for the superb Bolivian specimens.

Minerals of the pyrolusite
group   Plate 21

This group comprises several minerals, whose basic composi-tion is near MnO_2, but may, in fact, be rather complicated compounds:

PLATE 30   *Phosphate minerals*
Above Prismatic apatite crystals, combination of prism, pyramid and basal pinacoid. Snarum, Norway. *Center left* Tabular apatite crystas from an alpine fissure. Zillertal Alps, Tirol, Austria. *Center right* Six-sided prisms of pyromorphite partly coated with a thin layer of galena. Bad Ems, Germany. *Below left* Clear apatite crystal. Cerro de Mercado, Durango, Mexico. *Below center* Turquoise, polished.

Beta-MnO$_2$	pyrolusite, polianite	tetragonal
K$_2$Mn$_8$O$_{16}$	cryptomelane	tetragonal and monoclinic
Ba$_2$Mn$_8$O$_{16}$	hollandite	tetragonal and monoclinic
Pb$_2$Mn$_8$O$_{16}$	coronadite	tetragonal
(Ba, H$_2$O)$_2$ Mn$_5$O$_{10}$	psilomelane	monoclinic

Wads are water-bearing manganese oxides, but may also contain manganite. Pyrolusite and cryptomelane are the only common ones; 'psilomelane' found in collections is usually cryptomelane. Only pyrolusite can be identified easily and the would-be collector will find it best to keep all other varieties under the heading 'manganese dioxides'. For an accurate determination, chemical and X-ray methods must be used.

Pyrolusite, in radiating needle-like pseudomorphs after manganite, or granular, and its rare variety polianite are the only well crystallized minerals. All other manganese dioxide minerals occur as dense, black, smooth masses (psilomelane), stalactites, porous, frothy, earthy masses (wad), oolites, dendritic aggregates, where even in the tiniest grains no crystalline structure can be detected. Pyrolusite is black with metallic luster, and black streak. Its hardness is 5–6, but can hardly be determined on the very brittle and easily disintegrating mineral. The black color and black to dark-brown streak are characteristic for all members of the pyrolusite group. Wad has a browner streak. The hardness varies between 1 and 6 depending on compactness. All massive varieties are dull, but the surface of psilomelane may be shiny. Psilomelane comes from the Greek *psilos*, bald, and *melas*, black; pyrolusite from *pyr*, fire, and *lysis,* washing, alluding to its ability to remove green coloration from glass. The name 'lapis manganese', coined in the sixteenth century, is derived from the Greek *manganizein,* to clean.

Manganese metal was not discovered until 1774, and about the middle of the last century it began to be used in steels, which is still its main use. Pyrolusite is still used for clarifying glass, and also for making dry batteries, and many chemicals.

Formation and occurrence. MnO_2 occasionally occurs in hydro-thermal veins associated with other manganese minerals. Its chief occurrence, however, is in sedimentary deposits. Manganese oolites and manganese shales are marine deposits. Pyrolusite may also be formed by the weathering of other Mn-bearing rocks and deposits and may sometimes, after other constituents are carried away, be residual, forming greatly enriched masses. Siderite frequently contains rhodochrosite, which readily alters to pyrolusite. MnO_2 minerals are very widespread, so only a few localities are named.

GERMANY – Waldalgesheim, near Bingen, Rhineland, Fe-Mn deposit with pyrolusite: Göpfersgrün, in the Fichtelgebirge, Bavaria, dendritic MnO_2 in soapstone (steatite); also fine dendrites in the lithographic shales at Solenhofen, Franconia. AUSTRIA – Hüttenberg, Carinthia, in gossan, also at Erzberg, Styria. Also Postmasburg, South Africa; Brazil; India; Ghana; USSR.

WOLFRAMITE   Plate 19

CHEMISTRY $(Mn, Fe)WO_4$ is an oxide of tungsten and Fe or Mn respectively in a completely isomorphous series: $MnWO_4 - FeWO_4$, the Mn end-member being huebnerite and the Fe end-member being ferberite.

STREAK yellow-brown to black   MONOCLINIC   H $5–5\frac{1}{2}$
   SG 7.3

COLOR dark brown to black.

LUSTER resinous to sub-metallic (imperfect adamantine).

PROPERTIES perfect cleavage, fracture uneven, brittle; opaque.

CRYSTALS uncommon, thickly tabular, usually from roughly radiating to foliated, irregular cleavage, frequently associated with quartz.

IDENTIFICATION brown-black streak, weight, cleavage, crystal shape.

Wolframite affected the smelting of tin adversely; the old Saxon miners probably intended to express this by the name 'wolf'. Tungsten metal was discovered in 1781 by the chemist Scheele in scheelite. Tungsten is an important constituent of steels and other alloys, and is the filament in electric-light bulbs. Tungsten carbide is extensively used for the hard-cutting tips of tools.

Formation and occurrence. Wolframite and scheelite are associated with the formation of cassiterite (see p. 143) but may also be found in skarn ores and stibnite veins.

PORTUGAL – magnificent tabular, striated crystals (!!!), the world's finest, have recently been produced at Panasqueiros near Fundão associated with arsenopyrite, pyrite, cassiterite, apatite and siderite; some crystals reach 10 cm. in length and are of exceptional luster and perfection. GERMANY – fine crystals (!!–!!!) from Zinnwald, Erzgebirge; Neudorf, Harz. BOLIVIA – large crystals (!!!) at Llallagua and in other tin deposits. USA – bladed brownish crystals (huebnerite) in quartz matrix from several localities in Boulder, Gunnison, and Park counties, Colorado; free-standing prismatic crystals, sharp and clean, from the Alma area, Park Co.

COLUMBITE is the name of the isomorphous series between *columbite* (Fe, Mn)Nb_2O_6, and *tantalite* (Fe, Mn)Ta_2O_6. Orthorhombic. Limited to granite pegmatites, where it is found in black bladed to tabular crystals, sometimes exceeding 10 kg. in weight. Excellent crystals have been found in the pegmatite areas of Brazil, USA, Australia etc. Nb and Ta are important constituents of specialized steels.

URANINITE (Pitchblende)   Plate 20

CHEMISTRY fresh crystals (uraninite) UO_2, when decomposition sets in (pitchblende) UO_2 to U_3O_8; always with some thorium etc., lead and radium from radioactive decay (3.4 gm. per 10 tons uranium).

STREAK brownish-black, shiny   CUBIC   H 4–6   SG 9–10.5

COLOR pitch-black, brownish, greyish.

LUSTER semi-metallic, also dull or pitchy, especially on fractures.

PROPERTIES no cleavage, conchoidal fracture, brittle.

CRYSTALS uraninite, predominantly cubes, modified by octahedrons and rhombic dodecahedrons; massive and reniform (pitchblende), powdery.

PLATE 31   *Garnet*
Rhombic dodecahedrons of almandite in chlorite-mica schist. Zillertal Alps, Tirol, Austria. Beside it, perfect trapezohedron of almandite from Rhodesia.

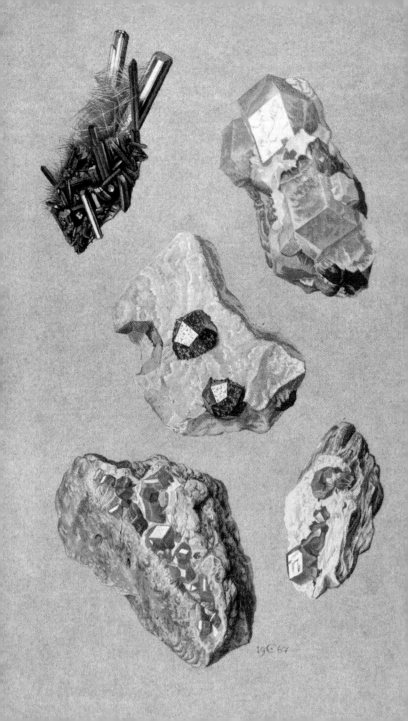

Identification pitchy luster, high specific gravity, radio-activity, very often associated with vividly colored secondary minerals.

Eighteenth-century Saxon miners gave the name of pitch-blende to a noticeably heavy but useless mineral. In 1789 Klaproth discovered in it a new metal, which he named uranium, but it did not prove possible to isolate this in a pure state until 1841. In 1896 Becquerel discovered the radio-activity of uranium, and two years later Marie and Pierre Curie discovered radium, after having used up about 1,000 kg. of pitchblende from Joachimstal. In 1938–39 Hahn and Strassmann succeeded in splitting the uranium atom.

During the last century uranium was used in the ceramic and glass industries for coloring purposes. During the first half of the twentieth century uranium ores were mined only for their radium content; not until after 1940 did uranium begin to be used for nuclear fission. The rapidly increasing demand led to a 'uranium rush', followed by over-production (in Canada alone over 1,000 occurrences were discovered).

Secondary uranium minerals

URANOPHANE (uranotile)	$CaH_2(UO_2)_2(SiO_4)_2.5H_2O$
'URANIUM MICAS': autunite	$Ca(UO_2)_2(PO_4)_2.10H_2O$
torbernite	$Cu(UO_2)_2(PO_4)_2.10H_2O$
carnotite	$K_2(UO_2)_2(VO_4)_2.3H_2O$

Uranophane frequently occurs as a mass of yellow matted needles. Of the numerous uranium micas (micaceous cleavage) only three important ones need be named here. Sulphur-yellow *autunite* is tabular or flaky, and so is brilliant green *torbernite; carnotite* occurs as yellow grains. Autunite is strongly fluorescent under ultra-violet light.

Formation and occurrence. Uraninite rarely occurs in granite pegmatites but is widespread in the hydrothermal zone (Co-

Plate 32 *Garnet and epidote*

Above left Epidote with finely fibrous actinolite. Knappenwand, Salzburg, Austria. *Above right* Pink grossular with idocrase. Lake Jaco, Chihuahua, Mexico. *Center* Brown garnets in quartzite. Gothenburg, Sweden. *Below left* Crystals of grossular. Val Maighels, Switzerland. *Below right* Demantoid. Val Malenco, N. Italy.

Ni-Ag-U veins, fluorite veins, copper-uranium and quartz-siderite veins). In the oxidation zone, and as a result of other decomposition processes, secondary uranium minerals occur as alteration products of pitchblende. These occur especially in sedimentary uranium deposits. Pitchblende may also be found in sediments – in fossil plants, for instance. This may mean appreciable uranium content in some coals, in natural oil and in bituminous rocks.

GERMANY – Kinzigtal and Krunkenbachtal in the Black Forest, pitchblende and secondary uranium minerals; Fuchs-bau, Fichtelgebirge, Bavaria, torbernite in granite; Schneeberg, Aue and Johanngeorgenstadt in the Erzgebirge, Co-Ni-Ag-U veins. CZECHOSLOVAKIA – Jachymov, classic pitchblende locality (monopoly until 1914). CONGO (Kinshasa) – Shinko-lobwe, Cu-U veins with uranophane and uranium micas, during World War II of great importance, now exhausted. This locality has provided handsome specimens of microcrystals of this species as well as other radioactive minerals. CANADA – Echo Bay, Great Bear Lake, abandoned workings as in the Erzge-birge; Blind River on Lake Huron, Ontario; fossil gravels. USA – Colorado Plateau, carnotite-mineralized sandstones; produces nearly half of the western world's uranium; magnifi-cent autunite crystals (!!!) at Daybreak mine, near Spokane, Washington. ENGLAND – beautiful reticulated masses of free-standing crystals (!!!) of torbernite have been found in Cornish mines.

Iron hydroxide group

goethite	alpha-FeOOH	orthorhombic, acicular
limonite	alpha-FeOOH + water	(impure goethite)
lepidocrocite	gamma-FeOOH	orthorhombic, platy
hydrolepidocrocite	gamma-FeOOH + water	
siderogel	FeOOH + water	amorphous

Whereas goethite and lepidocrocite represent two modifica-tions of FeOOH, limonite and hydrolepidocrocite contain varying amounts of water. Limonite has world-wide distribution, all others are uncommon.

GOETHITE and LIMONITE

CHEMISTRY little Mn replacing Fe; other impurities (e.g. SiO_2, P, Ba, V, Al, Ca) are admixed. Pure goethite: about 80 per cent Fe_2O_3 and 10 per cent H_2O; limonite: up to 14 per cent H_2O.

Goethite

STREAK brown   ORTHORHOMBIC   H 5   SG 4.3
COLOR brownish-black.
LUSTER sub-metallic, fibrous aggregates silky.
PROPERTIES perfect cleavage in one direction. Crystals acicular to prismatic, radiating, also in minute velvety needles.

Limonite

STREAK brown   H1–4 depending on compactness of aggregates   SG 3–4
COLOR yellow to brownish black.
LUSTER usually dull, but often with a lacquer-like black shiny coating.
PROPERTIES conchoidal or splintery fracture, massive and compact, partly radiating, stalactitic, cellular-porous, earthy as ochre, massive as common limonite, black and compact as bog iron ore (contains siderogel), in concretions as pea ore, oolites, and dendritic. Pseudomorphs after pyrite and other minerals are not uncommon.
IDENTIFICATION limonite is nearly always easily recognized, and only very dark smooth masses may be confused with MnO_2 (psilomelane); distinguished by brown and black streaks respectively.

Older names: xanthosiderite for yellow, and stilpnosiderite for black limonite, also pyrrhosiderite, pitch iron ore, iron glance stone, hydrogoethite, etc. The name limonite is derived from the Latin *limus,* bog (bog iron ore). It is the second most important iron ore after hematite. The ochre-yellow and brown colors found in Neolithic cave paintings are due to limonite, and it was also the first iron ore to be smelted.

Formation and occurrence. Occasionally found as the most recent formation in hydrothermal veins, but the majority of limonite deposits are formed by the weathering of iron-bearing rocks and minerals, some at least via amorphous FeOOH which crystallizes after ageing (gossan, laterite, pea ores). Reniform, spherical formations are typical, with radiating internal structures. Since practically all rocks contain some Fe, limonite, as coloring matter at least, is to be found everywhere. Bog iron ores (precipitated from lakes and marshes) and oolites are true sediments.

Lepidocrocite This rarer modification of FeOOH is red to reddish-brown, translucent, with dark orange streak. The crystals are foliated. In all other respects it is similar to goethite, also in occurrence, as in the Siegerland iron veins of Germany, where crystals of lepidocrocite may grow on top of brown limonite. *Hydrolepidocrocite* cannot be distinguished from limonite without special tests; both may occur together. *Siderogel* may sometimes be found in bog iron ore. *Manganite,* MnOOH is orthorhombic and crystallizes in prisms. Compact and earthy varieties have a brown streak, but may only be distinguished with some certainty from the MnO_2 minerals by chemical and X-ray methods.

Bauxite and the aluminum hydroxides

Bauxite is not a mineral, but a mixture of iron and aluminum hydroxides with clay minerals, quartz, hematite. The more important aluminum hydroxide minerals are:

diaspore	alpha-AlOOH	orthorhombic
boehmite	gamma-AlOOH	orthorhombic
gibbsite	gamma-Al(OH)$_3$	monoclinic

Bauxite, named after the French town Les Baux, is a weathering product of rocks, rich in aluminosilicates but containing very little free quartz, as argillaceous limestones mainly, but also nepheline syenites. During weathering in a warm climate only iron and aluminum oxides are left behind, all other constituents are carried away. Bauxite is usually whitish, but sometimes it may be red.

Bauxite is the only economic aluminum ore. Aluminum was discovered about the middle of the eighteenth century, but was not isolated until 1827 and a commercially useful smelting process did not get under way until 1866. Today aluminum is of the utmost importance in the aviation and engineering industries, in the building industry, for domestic purposes etc.

Class 5 Carbonates

Minerals consisting of one or more metals combined with the acid radical CO_3 belong here. The chemical formulae of hydrous carbonates contain (OH).

Calcite series

Minerals belonging to this group are hexagonal (trigonal), with perfect, easily developed rhombohedral cleavage, and brittle; their luster is vitreous, sometimes with a pearly appearance on cleavage planes.

	H	SG	cleavage	solubility in cold HCl
magnesite $MgCO_3$	4–4½	3	perfect	none
smithsonite $ZnCO_3$	5	4.4	perfect	good
siderite (chalybite) $FeCO_3$	4–4½	3.8	perfect	poor
rhodochrosite $McCO_3$)	4	3.5	perfect	poor
calcite $CaCO_3$	3	2.7	perfect	very good

MAGNESITE

Between magnesite and siderite there is complete miscibility. Mg-rich members: magnesite has Mg:Fe = 9:1; breunnerite Mg:Fe = 8:2; mesitite Mg:Fe = 6:4. Translucent to opaque, white to brown (mesitite is brown to black), with a white streak. Crystals are nearly always intergrown rhombohedra.

Magnesite is supposed to have been named after the region of Magnesia in Asia Minor. Magnesium was first obtained in 1808.

Magnesite is much in demand for refractory magnesite and chrome-magnesite bricks. Mg metal is a constituent of certain alloys, such as electron (90 per cent Mg with Al, Zn, Mn, Cu, Si).

Formation and occurrence. Magnesite is formed by metasomatic processes from limestones and dolomite, magnesite gel by the decomposition of Mg-rich silicates as serpentine; minute crystals or nodules of magnesite may occasionally be found in sediments. Well-developed rhombohedra occur in metamorphic rocks, such as talc schists.

AUSTRIA – crystals (!!) at Oberdorf an der Laming in Styria, breunnerite crystals in talc schist at Gösselkopf near Mallnitz, Carinthia. BRAZIL – transparent rhombohedral crystals (!!!) at Serra das Eguas, Bahia.

SMITHSONITE

CHEMISTRY $ZnCO_3$, with considerable manganese and iron at times, also some calcium and magnesium.

STREAK colorless.

COLOR mostly yellowish, greenish, bluish, sometimes vivid; also colorless, white, grey; commonly color-banded massive.

LUSTER waxy to vitreous in crystals.

PROPERTIES perfect cleavages, though seldom very apparent in usual massive material; brittle; translucent to transparent.

CRYSTALS rare, as low to steep rhombohedra; usually botryoidal massive, fibrous, crusts.

IDENTIFICATION crustal formation, heaviness, colors; dissolves in hydrochloric acid with bubbling.

Smithsonite is named after James Smithson, founder of the Smithsonian Institution in Washington, D.C.; the name *calamine* is sometimes given but was more correctly applied to hemimorphite.

Formation and occurrence. Formed as replacement mineral in oxidation zone of sphalerite deposits, particularly if dolomite or calcite are present. Frequently associated with galena, cerussite, calcite, dolomite, sphalerite, and limonite.

SOUTH-WEST AFRICA – large crystals (!!!) in yellow, green, pink in copper mines of Tsumeb area; also beautiful fine green crusts suitable for gems. ITALY – formerly in yellow stalactitic masses (!!) of considerable beauty at Iglesias, Sardinia (stalactites to 20 cm. diameter). GREECE – beautiful botryoidal and stalactitic material in greens, blues, etc., from ancient Athenian mines at Laurion, SE. of Athens. AUSTRALIA – large yellowish transparent rhombs (!!!) from Broken Hill, New South Wales. USA – beautiful blue-green massive crusts (!!!) at Kelly, near Magdalena, Socorro Co., New Mexico and once called 'bonamite'.

SIDERITE   Plate 24, also 6

CHEMISTRY usually with Mn, also Ca; forms solid solution with magnesite. Siderite covers a Mg:Fe ratio of 1:9 to 0:10.

STREAK white to pale brown.

COLOR yellowish-white, yellow to brown, grey, black.

LUSTER vitreous to waxy or oily; sometimes metallic, iridescent.

PROPERTIES perfect rhombohedral cleavage; brittle; transparent to opaque.

CRYSTALS common as low rhombohedrons, sometimes so flat as to be discoidal and with serrated edges; also coarse granular massive.

IDENTIFICATION cleavage, color and crystal forms.

Siderite comes from the Greek *sideros,* iron, and is otherwise known as chalybite. Britain's importance as a steel-producer was based on clay ironstone (impure siderite) occurrences. Siderite is an important iron ore today.

Formation and occurrence. Siderite is widespread in the hydrothermal zone, either in veins by itself or together with sulphide ores (Plate 6), accompanied by chalcopyrite, rhodochrosite or hematite. Extensive replacement deposits are known in which dolomite and calcite are replaced by siderite. Siderite also often occurs in alpine fissures, together with rutile. Sedimentary siderite occurs as clay ironstone, oolite or nodular iron ore.

BRAZIL – greenish-brown crystals (!!!) with quartz at Morro Velho, Minas Gerais. BOLIVIA – large brown crystals (!!) from Colavi. USA – tan to brown crystals in mines of Gilman district, Eagle Co., Colorado; stalactitic at Bisbee, Cochise Co., Arizona. GREENLAND – brown cleavage rhombs to 30 cm. in the cryolite deposit, Ivigtut, Arsuk Fjord. ENGLAND – beautifully crystallized in Cornwall, e.g. Camborne, Redruth, St

Austell. GERMANY – crystals (!!!) Lintorf, Hanover and Neudorf, Harz. AUSTRIA – fine crystals (!!!) at Erzberg, Styria.
ITALY – crystals (!!!) at Brosso, Traversella, Piedmont. FRANCE
– transparent crystals (!!) at Allevard, Isère. PORTUGAL – tan
discoidal crystals, sometimes clear inside, at Panasqueiros.

RHODOCHROSITE

CHEMISTRY $MnCO_3$ manganese carbonate; a complete series
extends to siderite and to calcite; also commonly contains
iron.

STREAK colorless   HEXAGONAL (trigonal)   H $3\frac{1}{2}$–4   SG
3.5

COLOR usually pale to deep rose red, also brownish red, greyish,
brownish, greenish.

LUSTER vitreous; often pearly on cleavage planes.

PROPERTIES perfect rhombohedral cleavage; brittle; translucent to transparent.

CRYSTALS simple rhombohedrons, sometimes rough-surfaced
due to mosaic structure; also coarse to fine granular; banded.

IDENTIFICATION distinctive pink color and banding; cleavage;
dissolves in warm HCl with bubbling.

Formation and occurrence. In hydrothermal veins with other
carbonates, sulphides, quartz, etc.; secondary mineral in Mn
deposits; rarely in pegmatites. USA – magnificent deep pink
rhombs (!!!) on matrix with drusy quartz from Colorado mines:
John Reed mine, Lake Co., Sweet Home mine, Park Co.; also in
mines in Chaffee, Gilpin, and Saguache counties; rhombs reach
10 cm. on edge; some are clear and can be cut into faceted gems.
Sharp pink but only translucent to nearly opaque rhombohedrons occur in the mines of Butte, Silver Bow Co., Montana.
ARGENTINA – beautiful stalactites (!!!) handsomely banded,
from Capillitas, Catamarca Province, cut into ornaments.
GERMANY – fine specimens (!!) in Grube Wolf at Herdorf,

PLATE 33   *Olivine, kyanite, staurolite, andalusite*
Above left Olivine bomb with spinel between olivine grains. Dreiser Weihe,
Eifel, Germany. *Above right* Crystal of olivine. *Center* Kyanite and staurolite in
paragonite. Two of the blue kyanite crystals are regularly intergrown with
staurolites. Sponda Alp, Pizzo Forno, Switzerland. *Below left* Interpenetrant
twin of staurolite. Brittany, France. *Below right* Crystal of andalusite, or
'chiastolite', polished; regularly intergrown impurities form a cruciform pattern.

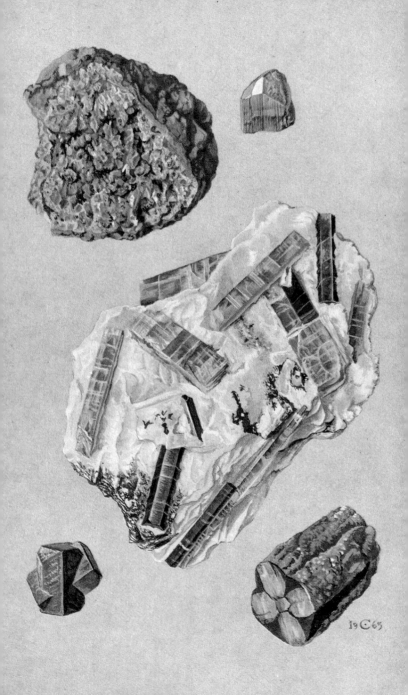

Rhineland-Westphalia, also at Beiersdorf. ROMANIA, YUGO-
SLAVIA – pale pink crystals at Kapnik and Trepča.

CALCITE   Plates 22, 23, 26

CHEMISTRY $CaCO_3$, calcium carbonate, forms series to siderite
and rhodochrosite, but usually very little Mg, Fe, Mn, or
more rarely Zn, Ba, Sr, Pb.

STREAK white, but may also be colored in very impure varieties
    HEXAGONAL (trigonal)   H 3   SG 2.7

COLOR colorless, white, and other pale shades, as yellow,
brown, purplish, etc.

LUSTER vitreous.

PROPERTIES perfect rhombohedral cleavage; brittle; transparent, rarely opaque.

CRYSTALS all sizes are common. Flat, steep or very steep
rhombohedrons (the cleavage rhombohedron, however, is
very rarely developed), scalenohedrons, prisms, and basal
planes are the most frequent forms. The habit of calcite is
more varied than that of any other mineral. Twinning (see
Pl. 22, 23 and Fig. 20) is common, on the basal plane, or on
rhombohedrons (butterfly twins); lamellar twins are also
common. Massive as coarse to fine granular, or compact
aggregates, spiky, fibrous, earthy. As pseudomorphs after
various minerals.

IDENTIFICATION by crystal form; the normal rhombohedron
(= cleavage rhomb) is rare (predominant in all the other trigonal carbonates!); by perfect cleavage; effervesces in dilute
hydrochloric acid (1 part HCl to 3 parts water).

In 1669 Bartholin discovered the phenomenon of double refraction on 'Iceland spar', a particularly clear variety of calcite
from Iceland. Many crystals have this property of splitting an
incident ray of light into two differently refracting rays. In
calcite the difference in refraction is particularly large, so that
any writing will appear to be quite clearly doubled (Pl. 22).

PLATE 34   *Topaz*
Above Single crystal, showing several prisms and pyramids. Nerchinsk,
USSR. *Center* Crystals in tourmaline rock, with quartz. Schneckenstein,
Vogtland, Germany. *Below left* Crystal consisting of prism and pyramid. Ouro
Preto, Minas Gerais, Brazil. *Below right* Blue topaz, 'envelope' shape. Minas
Gerais, Brazil.

1 *Limestone and marble.* Compact to granular masses of calcite, pure or with admixed clay, sand, hornstone, etc., are known as *limestone. Marble* is a metamorphosed limestone with uniform, large or small grain size. For technical purposes, any limestone suitable for cutting and polishing is called 'marble'. Special forms are: *calcareous tufa* deposited from cold, CO_2-rich water; *travertine, deposited* from warm water solution; *chalk,* consisting almost entirely of the shells of microscopic marine organisms; *oolites,* consisting of spherical-foliated structures of 1–5 mm. diameter; bituminous limestone; loess, wind-transported limestone dust.

2 *Fissure fillings and sinter formations.* Crevices, cracks and fissures in limestones and calcareous sandstones may often fill with granular or fibrous calcite. Calcite formations in caves include: encrusted sinters on walls, stalactites (hanging down from above), stalagmites (growing up from the ground); occasionally also mountain milk (earthy) and agaric mineral (moon milk), which under natural damp conditions are earthy and soft, but hard and very light when dry; very rarely *cave-pearls,* formed in hollows by circular water movement.

3 *Organic limestone formations.* The skeletons and shells of many living creatures consist of calcite (originally aragonite); other organisms encrust themselves with lime.

Calcite has many uses. Iceland spar is extremely valuable for optical instruments. Colorful, fibrous varieties are occasionally cut and polished into ornaments, those with multi-colored layers being erroneously called 'onyx'. White, uniform marble is prized for sculpture and architecture. One of the most famous buildings in the world, the Parthenon on the Acropolis in Athens, was built (447–438 BC) from marble from Mount Pentelikon. Enormous quantities of ordinary limestone are used as building and ornamental stones; by the chemical industry in lime kilns for cement-making; as fertilizers in agriculture; and as a flux in metallurgy.

Formation and occurrence. Good crystals in a variety of different habits are abundant in hydrothermal ore veins, in alpine fissures, and filling druses of basalts and similar rocks. Fissure fillings and all kinds of sinter formations are mostly deposited from cold solutions. Huge masses of calcite occur in sediments, as limestone alone, or mixed with clay as *marl,* or as the

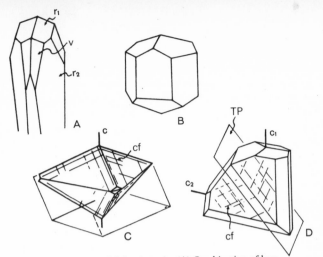

20 Calcite crystals and dolomite twin. (A) Combination of low rhombohedron r_1, steep rhombohedron r_2, and scalenohedron v. The specimen in Pl. 22 displays such forms. (B) Low rhombohedron and prism as in Pl. 23, lower left. (C) Two rhombohedrons and prism twinned on the basal plane; at the top is shown the complete form (common in dolomite), cf cleavage fissures. (D) 'Butterfly' twin (see upper specimen, Pl. 22). c_1 is the axis of one individual, c_2 the corresponding axis of the other. TP is the twinning plane, cf cleavage fissures. (E) Combination of a low rhombohedron with very steep rhombohedron, common in calcite (Pl. 23, upper specimen).

cementing matter of other sediments, etc. Pure limestone is altered to marble by metamorphism; calcareous mica-schists are formed from sandy marls. Crystals may be expected to occur practically everywhere that limestones, dolomites or marbles occur, as well as in various ores or mineral deposits. Only occurrences of particularly beautiful crystals, or otherwise notable localities, are given.

ENGLAND – superb specimens (!!!) from numerous localities, classical and world renowned, especially from Egremont and Frizington, Cumberland; Eyam and Ecton, Derbyshire; Stank mine, Lancashire; 'nailhead spar' crystals (!!) from Alston Moor, Cumberland. ICELAND – enormous clear masses and crystals (one single crystal measured 6 m. × 2 m.!) in cavities in basalt at Helgustadir, Eskifjord. FRANCE – curious calcite crystals apparently filled with pale grey sand ('sand calcite') at

Fontainebleau near Paris. GERMANY – crystals (!!–!!!) St Andreasberg, Harz and other mines here and in Saxony, etc. SWITZERLAND – common in the Jura and in many alpine vugs. AUSTRIA – crystals (!!) at Bleiberg, Carinthia. USA – yellow to colorless crystals in trap rock quarries of Connecticut and New Jersey; splendid transparent and brilliant crystals enclosing bright-red native copper (!!!) in mines of Keeweenaw Peninsula, Michigan; honey-colored crystals, also purplish, as small to large scalenohedrons to 60 cm. in lead-zinc mines of the Tri-State district around the junction of Kansas, Missouri, and Oklahoma; pale brown crystals (!!) on tan sandstone in Jennings Co., Indiana; fine 'sand calcites' (!!!) at Rattlesnake Butte, Washington Co., South Dakota; spectacular brilliant orange-red crystals (!!!) have been found in copper deposits of Bisbee, Cochise Co., Arizona and owe their coloration to abundant inclusions of hair-like cuprite (var. *chalcotrichite*); nailhead crystals (!!) with mountain leather, Metaline Falls, Pend o'Reille Co., Idaho. MEXICO – fine tabular crystals (!!), sometimes amethystine, from silver mines of Guanajuato; in great variety of forms, especially large tabular crystals (!!!) from Charcas, San Luis Potosi; abundant as rhombs (!) in Naica, Chihuahua and elsewhere. SOUTH-WEST AFRICA – rhomb crystals at Tsumeb and Grootgontein.

Calcite onyxes occur in many countries, notable deposits being in Argentina, a fine green, translucent material, in Mexico, Algeria, France, etc.

DOLOMITE and ANKERITE   Plate 24

CHEMISTRY dolomite, $CaMg(CO_3)_2$, ankerite, $Ca(Fe, Mg)(CO_3)_2$. Excess of Ca due to intergrowth with calcite. A complete series exists between dolomite and ankerite.

STREAK white   HEXAGONAL (trigonal), but with lower symmetry than calcite   H $3\frac{1}{2}$–4   SG dolomite 2.9, ankerite 2.9–3.8, depending on Fe content.

COLOR dolomite crystals colorless, white and impure, mostly yellowish, compact varieties mostly white, brownish-grey to

PLATE 35   *Sphene and zircon*
Above twinned crystal of sphene. *Center* Sphene crystals in fine-grained chlorite and foliated calcite. Habachtal, Salzburg, Austria. *Below* Crystal of zircon, a combination of prism and bi-pyramid.

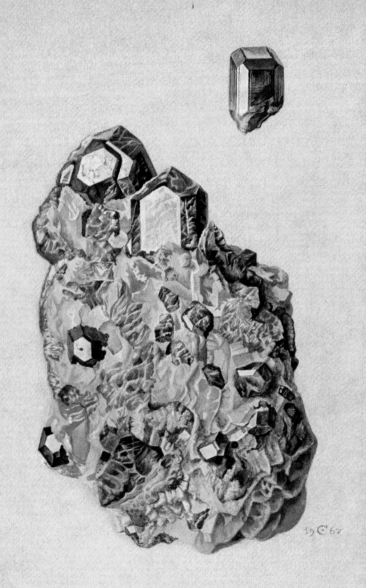

black, also differently colored; ankerite is usually white to yellowish-brown.

LUSTER vitreous.

PROPERTIES perfect rhombohedral cleavage, brittle, transparent to opaque.

CRYSTALS crystals of dolomite are common, ankerite crystals are rather rare. The cleavage rhomb is the most common form, often combined with the basal plane; crystals with many faces are rare. Twins on the basal plane are common, lamellar twinning rarer than in calcite. Granular or compact, cellular-porous; weathered specimens may sometimes be friable.

IDENTIFICATION differences from calcite: dolomite does not effervesce in dilute HCl; cleavage rhombohedra prominent. Difficult to distinguish tiny crystals and grains from magnesite, siderite, and especially from ankerite, but these carbonates are relatively rare in sediments.

Saussure named dolomite after Déodat Dolomieu, who in 1791 was the first to notice the difference between this mineral and calcite. Ankerite is named after the Austrian mineralogist M. J. Anker. Dolomite is also known as bitterspar, ankerite as brown spar, but these names have other meanings: bitterspar is also used for magnesite, and brown spar for manganiferous ankerite, and for breunnerite. Dolomite rock is used as a building stone, for making refractory bricks, and for the production of magnesium.

Formation and occurrence. Dolomite occurs in hydrothermal ore veins, but less often than calcite. Dolomite, and especially ankerite, may also be formed metasomatically by the alteration of limestones, as in the Erzberg of Styria and Carinthia. In sediments dolomite occurs like limestone, but its formation is still not completely understood. Some dolomite, but by no means all, is formed from limestone with the addition of Mg during diagenesis. Dolomite rocks are metamorphosed, similarly to calcite, into dolomite marbles. Good crystals may often be found in talc-schists, serpentinites, or similar rocks.

PLATE 36 *Idocrase (vesuvianite)*
Brown idocrase crystals, combinations of two prisms and a pyramid, in blue calcite. Val di Fassa, Alto Adige, Italy. *Above* Single crystal with pyramid flattened by basal pinacoid. Monte Somma, Vesuvius, Italy.

Only places where good crystals occur are given. Unless ankerite is specially mentioned, the localities refer to dolomite only.

SPAIN – clear rhombs to 5 cm. on matrix (!!!) near Eugui, Navarra; dark crystals *(teruelite)* in Barranca del Salobral, Teruel. AUSTRIA – crystals (!!!) at Oberdorf a.d. Laming, Styria, also at the Erzberg and Sunk near Trieben; Leogang, Salzburg. SWITZERLAND – clear crystals in marble (!!) at Binnatal and at Campolungo (!!). GERMANY – small glassy rhombs with chalcopyrite at St Goar, Rhineland; in the siderite veins of the Eupel and Neue Hardt deposits in Siegerland; crystals (!!) Freiberg, Saxony (Mn-ankerite); with fluorite at Wölsendorf, Bavaria. ENGLAND – choice crystallizations (!!!) from Cumberland and less commonly Cornwall mines. YUGOSLAVIA – crystals (!!!) at Trepča. ITALY – crystals (!!!) Traversella and Brosso, Piedmont (crystals to 5 cm), and in the Pfischtal, Trentino. USA – small pink crystals with chalcopyrite in lead mines of Tri-State area; similar crystals in the cavities of the Niagara limestone in the Lake Erie, Ontario, and Huron region; rhombohedral crystals in emerald-bearing vugs in gneiss around Hiddenite, Alexander Co., North Carolina; discoidal crystals (!) with quartz Carson Hill, Calaveras Co., California. MEXICO – common in lead-zinc-silver mines of Chihuahua and other states.

Aragonite group

Members of this second carbonate group are orthorhombic. The cleavage in two directions is much less distinct than in hexagonal carbonates. Granular aggregates are never as obviously coarsely cleavable as calcite or siderite. The crystals are brittle, with conchoidal fractures. $CaCO_3$ is the only carbonate to occur in two modifications and it is therefore a member of both groups.

		H	SG	Cleavage
aragonite	$CaCO_3$	$3\frac{1}{2}$–4	2.9	indistinct
strontianite	$SrCO_3$	$3\frac{1}{2}$	3.7	distinct
witherite	$BaCO_3$	$3\frac{1}{2}$	4.3	indistinct
cerussite	$PbCO_3$	3–$3\frac{1}{2}$	6.5	indistinct

ARAGONITE   Plate 24

CHEMISTRY some strontium substitutes for calcium, and rarely lead.

STREAK white.

COLOR colorless, white, yellowish, reddish, green, bluish, grey, transparent to cloudy.

LUSTER vitreous.

CRYSTALS prismatic with wedge-shaped terminations, often apparently hexagonal due to twinning, steep pyramids. Radiating, spiky, or fibrous, stalactitic, sinter crusts, botryoidal.

IDENTIFICATION soluble in cold dilute acid with effervescence (different from strontianite); on boiling, powdered aragonite in a cobalt solution is stained violet, calcite remains white; pseudohexagonal twins common.

Its name is derived from the locality Molina de Aragón, Spain, where fine twin crystals occur in red clays. Arborescent 'flos ferri' is a weathering product of siderite. Aragonite is also deposited from hot springs, and is the chief constituent of pearls formed in marine shells, which consist of layered aragonite crystal prisms and a horny substance. The 'mother of pearl' lining the interior of shells is the same mineral.

Formation and occurrence. Aragonite is deposited mainly from low-temperature solutions, usually as the last hydrothermal product; thus it occurs in druses of basaltic rocks, or in lavas together with zeolites. Aragonite is often a weathering product of ultrabasic and magnesium-rich rocks and of siderite. In sediments it is associated with clays, gypsum, sulphur, halite etc. Finally the shells and solid parts of snails, coral polyps, mussels and other molluscs are wholly or partly aragonite.

SPAIN – sharp pseudohexagonal twins (!!!) stained red but also nearly colorless or sometimes faintly violet at Molina de Aragón, to 5 cm. diameter and 8 cm. long. ITALY – superb crystal groups (!!!) with sulphur and celestite in the sulphur mines of Sicily at Racalmuto, Cianciana and Agrigento to 8 cm. diameter. CZECHOSLOVAKIA – long prisms, often clear (!!) at Bilina in basalt cavities. HUNGARY – fine white to clear colorless twin crystals (!!!) from Herrengrund and Dognacska. AUSTRIA – beautiful flos ferri (!!!) from Erzberg, near Eisenerz and at Hüttenberg-Lölling, Carinthia; also crystals (!!) at Werfen and Leogang, Salzburg. POLAND – lead-bearing variety tarnowitzite (!!) at Tarnowitz, Silesia.

Aragonite twins are replaced by other minerals, as native copper in Corocoro, Bolivia, by calcite in clay at Las Animas, Bent Co., Colorado, etc.

STRONTIANITE Plate 24

Usually contains some calcium; named after the locality at Strontian, Scotland where it occurs in veins with galena and barite. Crystals rare and then as pseudohexagonal aggregates of numerous small crystals. Resembles aragonite in properties but distinguished by vivid red flame when powdered mineral is mixed with powdered borax and strongly ignited; the color is due to the strontium. GERMANY – good crystal aggregates at Drensteinfurt, Ascheburg and Ahlen, Westphalia. AUSTRIA – fine specimens (!!) from Leogang near Salzburg and Brixlegg.

WITHERITE

A fairly rare mineral, named after English mineralogist W. Withering. The typical white to pale tan-yellow groups are aggregates of many smaller crystals forming rounded hexagonal balls (Illinois) or tapered hexagonal crystal growths (England). Not easily distinguished when massive but powdered mineral with borax produces a green flame when ignited (try in a darkened room to see the green!). ENGLAND – fine specimens (!!) Fallowfield, Hexham in Northumberland to 8 cm. diameter; also from Alston Moor, Cumberland (!!) and Settlingstones mine, near Fourstones (!!). USA – large rude aggregates (!!!) to 10 cm. diameter from the fluorite veins of Rosiclare, Hardin Co., Illinois.

CERUSSITE

STREAK white.

COLOR colorless, white, grey, yellow, brown (due to limonite) black (due to galena).

PLATE 37 *Beryl*
Above left Beryl, variety emerald, embedded in mica schist. Habachtal, Salzburg, Austria. *Above right* Crystal of aquamarine. Nerchinsk, USSR. *Below* Common beryl in pegmatite, with rounded edges; the crystal has fractured and been re-cemented with quartz. Poběžovice, Czechoslovakia.

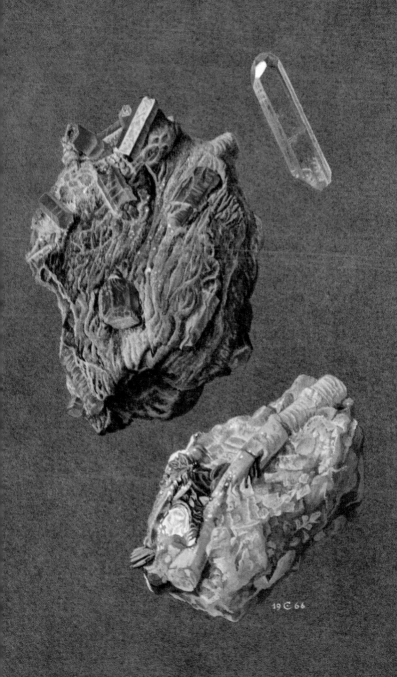

LUSTER adamantine.

CRYSTALS pseudo-hexagonal, tabular-platy or spiky, inter-penetrating cruciform twins frequent, botryoidal, radiating or clustered. Cerussite intermixed with limonite is nearly indistinguishable from it.

IDENTIFICATION crystals striking by their weight, luster and form (twinning), soluble in nitric acid with effervescence (very similar anglesite is insoluble).

Cerussite, from the Latin *cerussa,* white lead, is also called white-lead ore and is occasionally used in pigments.

Formation and occurrence. Limited to the oxidation zones of lead-zinc deposits; associated with galena, anglesite, limonite, smithsonite, and if the deposit also contains copper, with mala-chite and azurite.

SOUTH-WEST AFRICA – abundant and fine twinned crystals (!!!) from Tsumeb, some to 12 cm. diameter. AUSTRALIA – Broken Hill, New South Wales is noted for superb reticulated 'stars' or 'snowflakes' of twinned cerussite (!!!) to 20 cm. diameter, also large single crystals (!!!); in crystals (!!) at Dundas, Tasmania. ITALY – crystals (!!–!!!) from Monte Poni and Monte Vecchio, Iglesias, Sardinia. GERMANY – twinned crystals (!!) at Friedrichssegen, Ems. Also from metal mines in Austria, Czechoslovakia, USA, etc.

AZURITE   Plate 25

CHEMISTRY $Cu_3OH(CO_3)_2$, a copper hydroxy-carbonate.
STREAK sky-blue   MONOCLINIC   H $3\frac{1}{2}$–4   SG 3.8
COLOR dark blue (crystals) to pale blue (massive forms).
LUSTER vitreous to earthy in some compact types.
PROPERTIES distinct cleavage; uneven fracture, sometimes conchoidal; very brittle; transparent to translucent, also opaque.

PLATE 38   *Tourmaline*
Above left Crystal with vari-colored terminations. Mesa Grande, California, USA. *Above right* Large crystal with triangular cross-section. The curved appearance of the faces is due to the combination of several prisms. South-West Africa. *Below left* Schorl in feldspar with muscovite. Switzerland. *Below right* Blue tourmaline or indicolite. Minas Gerais, Brazil.

CRYSTALS common, short prismatic with striated sides, to thick tabular, edge-shaped, etc.; also in massive forms, veinlets, botryoidal, etc.

IDENTIFICATION dark blue color, crystals; usually associated with malachite; fairly heavy.

The following mineral names have been derived from *lazhward,* the Persian word for 'blue': azurite (via the French *azur,* blue); lapis lazuli, whose chief constituent is lazurite; and lazulite (see pp. 249 and 184). In old texts azurite has also been described as lazurite. Confusions were frequent; in Aristotle's *De Petris* the properties of azurite as well as those of lapis lazuli are attributed to what he names as *lazhward.* Theophrastos, on the other hand, gives an unambiguous description of lapis lazuli, although under the name of *sappheiros* (it is 'as if spotted with gold', and lapis lazuli often has inclusions of pyrite grains), and for azurite he mentions its metal content and its association with malachite.

Formation and occurrence. As for malachite (see below).

MALACHITE   Plate 25

CHEMISTRY $Cu_2(OH)_2(CO_3)_2$, copper hydroxy carbonate.
STREAK pale green   MONOCLINIC   H $3\frac{1}{2}$–4   SG 4
COLOR emerald to dark green; also pale green in massive types.
LUSTER vitreous, for radiating aggregates silky.
PROPERTIES cleavage distinct, fracture conchoidal, brittle, translucent to opaque.
CRYSTALS very rare, usually foliated, radiating fibrous aggregates, compact or earthy, stalactitic botryoidal, banded like agate. Pseudomorphs after azurite, cuprite, tennantite, tetrahedrite, calcite and others are not uncommon.
IDENTIFICATION color, association with azurite.

Malachite is Pliny's *molochites,* from the Greek *malache,* mallow. Copper could easily be obtained from malachite and azurite by smelting, so that during the Bronze Age these were the minerals principally quarried – particularly from Mt Sinai, which supplied Egypt with copper. Malachite has long been used as an ornamental and gem stone, and for seals. It was considered an antidote against, and cure for, malignant growths and, in medieval times, a talisman of malachite protected children from accidents. During the nineteenth century malachite was a popular ornamental stone; the large masses found in the Urals veneered table tops, vases, boxes and other objets d'art.

Formation and occurrence. Malachite and the somewhat rarer azurite are formed in the oxidation zones of copper deposits only in the presence of other carbonates. Associated minerals include limonite, cuprite, native copper, tenorite, pyrolusite, calcite, chrysocolla ($CuSiO_3.nH_2O$), smithsonite and cerussite.

USSR – Mednorudyansk, near Nizhny Tagil, and Gumeshevo in the Urals, SW. of Sverdlovsk, supplied the famous large slabs of malachite in the nineteenth century; from Mednorudyansk about 1850 came a slab about 6 m. long and weighing 50 tons. FRANCE – Chessy near Lyons, famous pseudomorphs of malachite after cuprite, very fine crystals of azurite *(chessylite)*. CONGO (Kinshasa) – Etoile du Congo mine, Katanga, malachite suitable for polishing. SW. AFRICA – Tsumeb, Otavi district, azurite and malachite (!!!), malachite often pseudomorphous after azurite with some crystals only partly altered. USA – Morenci, Greenlee Co., and Globe District, Pima Co., Arizona, crusts and stalactites with alternating layers of malachite and azurite (!!!), also fine azurite crystals (!!!). GERMANY – Friedrich Christian quarry near Schapach, Black Forest, good malachite. AUSTRALIA – Burra Burra, South Australia, polishable malachite.

Class 6 Sulphates, molybdates, tungstates

This class includes all compounds of metals with SO_4 (hydrated or anhydrous), MoO_4 or WO_4.

ANHYDRITE

CHEMISTRY $CaSO_4$, calcium sulphate.
STREAK white   ORTHORHOMBIC   H 3–4   SG 2.9
COLOR colorless, white, grey, bluish, reddish, black, purplish.
LUSTER vitreous.
PROPERTIES cleavage, perfect in one direction, good in other two, so that cubes with dissimilar faces are formed; fracture splintery, brittle, transparent to opaque.

CRYSTALS rare, cubic or prismatic. Mostly granular to compact. IDENTIFICATION cleavage; no effervescence with acids, even on warming; harder than gypsum.

During the eighteenth century the Abbé Nicholas Poda of Neuhaus discovered a new mineral in the salt workings at Hall in the Tirol, which he thought to be a chloride of calcium and therefore named it *muriacite*. In 1795 Klaproth recognized the true composition. Werner then added the name *Würfelspat* (cubic spar), but later he renamed it anhydrite (from the Greek *hydor*, water, and the prefix *an-*: 'without water'), in contrast to water-containing gypsum. Uses as for gypsum.

Formation and occurrence. Anhydrite occurs as an independent sediment mineral associated with calcite, clay and gypsum, often also in salt deposits. It occurs rarely in hydrothermal veins and in alpine fissures, more often in druses of basaltic rocks together with zeolites.

GERMANY – crystals in massive anhydrite (!), near Celle, Lower Saxony, with quartz and boracite crystals; rarely in crystals (!!) at St Andreasberg, Harz. SWITZERLAND – superb crystals (!!!) came from the Simplon Tunnel, reaching lengths of 20–30 cm. and of colorless to pale violet hue. MEXICO – recently as blue bladed crystals (!!) with calcite and sulphides from Naica, Chihuahua.

Barite group

Minerals belonging to this group are orthorhombic and cleave into lozenge-shaped tablets, because of their perfect basal, and somewhat less perfect prismatic cleavages. The crystals are often tabular parallel to the base, but may also be columnar (barite and particularly celestite) or pyramidal (anglesite). All and noticeably heavy and completely soluble only in concentrated sulphuric acid.

		H	SG	basal cleavage	luster
celestite	$SrSO_4$	3–3½	4	perfect	vitreous
barite	$BaSO_4$	3–3½	4.5	perfect	vitreous
anglesite	$PbSO_4$	3	6.3	less distinct	adamantine

CELESTITE   Plate 26

CHEMISTRY strontium sulphate, with some Ca, Ba.

STREAK white.

COLOR colorless, white, yellowish, commonly bluish or distinctly blue, more rarely greenish, reddish.

CRYSTALS quite common; also granular to roughly cleavable and nodular.

IDENTIFICATION cleavage; differs from carbonates in not effervescing in hot acids; differs from barite in coloring flame red.

Celestite from Pennsylvania was described in 1791 by Schütz as 'fibrous barite', and recognized as strontium sulphate by Klaproth in 1797. Werner named the mineral after the Latin *coelestis,* sky-blue. Strontium salts are used for making fireworks, and in the manufacture of special glasses and glazes, and strontium metal is used in special alloys.

Formation and occurrence. Like calcite, dolomite, and marls, celestite is formed in marine sediments; not so much, however, during sedimentation itself, as subsequently in druses and fissures from migrating (not hydrothermal) solutions. Occasionally crystals may be found in some fossil cavities, as in ammonites. Hydrothermal formation is rare. The chief associated mineral is calcite. GREAT BRITAIN – Yate near Bristol, deposit with fine crystals. ITALY – Agrigento, Sicily, crystals (!!!). MEXICO – Matehuala, San Luis Potosi, large (!!) crystals. SWITZERLAND – La Rechenette, near Biel in the Swiss Jura (Pl. 26), also ammonites with celestite crystals. AUSTRIA – Oberdorf a.d. Laming, Styria, crystals (!); Bleiberg, Carinthia, small blue tablets. GERMANY – Giershage, near Stadtberge, and elsewhere in Westphalia, crystals (!!). USA – enormous crystals (!!!) were once found in a large cave on South Bass Island, Put-In-Bay, Lake Erie; common as small sharp blue crystals in cavities in the Lockport dolomite around Lake Erie; fine blue tabular crystals (!!) in cavities in limestone at Clay Center, Ottawa Co., Ohio and in the surrounding region wherever quarries expose the formations; blue crystals (!) in geodes at Lampasas and Mount Bonell region, Texas; blue crystals lining cavities from near Emery, Emery Co., Utah.

BARITE   Plates 27 and 5

CHEMISTRY barium sulphate, often with some strontium.
STREAK white.
COLOR white, and in various, usually pale, colors; compact
    varieties may even be black owing to coal or bitumen in-
    clusions.
CRYSTALS common, often large and well developed (Pl. 27),
    thinly tabular to fan-shaped or as barite roses ('desert roses.),
    roughly cleavable to very compact, botryoidal, stalactitic.
    Transparent to translucent.
IDENTIFICATION crystal habit and cleavage, weight; compare
    celestite (p.175); green color in flame.

The name, from the Greek *barys,* heavy, only came into use
around 1800. 'Heavy spar' is an older name. In 1604 a Bologna
cobbler found that when a mineral found nearby was heated
together with combustible materials, a phosphorescent substance was formed. The 'Bologna spar', also known as 'Bologna
luminous stone', later proved to be barite. White barite pigments are now replaced by titanium whites (see p. 123). Large
quantities of barite are used as a filler, providing a smooth
finish for art papers and especially photographic papers, card-
board, and textiles. It is also needed by the leather industry
and in ceramics (enamels). By far the largest part of the total
production (2.6 million tons a year at present) goes to the oil
industry to weight the drill-mud. Radiation-proof buildings
are constructed from 'heavy concrete' made of barite debris.
    Formation and occurrence. Widespread with hydrothermal ores
of all kinds or alone in veins; larger, layered deposits of barite
are considered to be sedimentary. Associated minerals are:
sulphide ores, especially chalcopyrite and tetrahedrite-tennantite, also fluorite, quartz, calcite, Mn-minerals, siderite, hematite. Workable deposits are situated all over the world, and good
crystals are by no means rare.
    GERMANY – Christian-Levin quarry, near Essen-Borbeck,
Westphalia, large clear crystals; Clarashall quarry, near Baum-
holder, Rhineland-Palatinate, beautiful clear tablets, some including cinnabar (!!); Ilfeld, Harz, with manganite; Freiberg,
Saxony; Münzenberg, Hesse, barite roses in sandstone; Mün-

stertal, Black Forest, crystals; Wölsendorf, Bavaria, roughly spathic, and coarsely flaky (!) with fluorite. GREAT BRITAIN – The most beautiful known specimens come from the following localities: Egremont, Frizington, and Alston Moor, Cumberland (in 1851 a crystal weighing over 100 lb. was sent from Alston Moor to London for the Great Exhibition); Weardale, Durham; Dufton, Westmorland. ROMANIA – Baia Sprie, crystals with stibnite needle inclusions (!!). ITALY – Monte Paterno, near Bologna, radiating concretions with gypsum in marls, 'Bologna stone'. AUSTRIA – Bleiberg, Carinthia, parallel-foliated hemispheres; Hüttenberg, Carinthia; Brixlegg and Schwaz, Tirol, with tetrahedrite-tennantite (Pl. 5). SWITZERLAND – Binnental, Valais, good crystals. TAIWAN – Hokuto, Pb-bearing barite 'hokutolite'.

USA – fine greenish-blue tabular crystals (!!) on matrix near Sterling, Weld Co., Colorado (5–10 cm. long), also sharp, clear yellow-brown tabular crystals (!) at Gilman, Eagle Co; barite 'desert roses' often of singular beauty, in soil around Norman, Oklahoma; largely flawless tapered crystals to 15 cm. long, of yellow-brown color, and superb quality (!!!) in geodes in Pierre shales along Elk Creek, Mead Co., South Dakota; large tan, bladed crystals (!) along sea cliffs of Palos Verdes, Los Angeles Co., California. CANADA – crystals (!!) with fluorite, Grand Forks, British Columbia.

A favorite collection species because of the number of localities furnishing fine specimens; undoubtedly the finest quality in general appears in the superb crystals of the many English deposits.

ANGLESITE

Usually pure lead sulphate resembling cerussite closely (p. 168) and similarly formed. The name is derived from the Welsh island of Anglesey. Good crystals (!!) come from Monte Poni near Iglesias, Sardinia, from Tsumeb (!) and other lead deposits.

GYPSUM   Plate 28

CHEMISTRY $CaSO_4 \cdot 2H_2O$, hydrated calcium sulphate, often impure owing to admixed bitumen, coal or clay, hematite, etc.

21 Gypsum. (A) Side view of a crystal with its b-axis perpendicular to the plane of the paper. P_1, vertical, and P_2, inclined monoclinic prisms; b, a pair of pinacoids, to which the perfect cleavage of the mineral is parallel. The continuous lines on b indicate the second-best cleavage direction (a cracking sound is heard on breaking the crystal), and the broken lines show the least well-developed cleavage direction (breaks with a crunching sound). (B) Swallowtail twin. (C) Montmartre twin. The cleavage cracks in both types of twinning indicate the original orientation of individual crystals. The Montmartre twin is characteristically rounded (see Pl. 28).

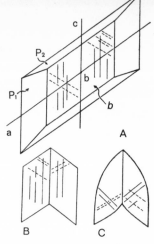

STREAK white   MONOCLINIC   H 2   SG 2.3

COLOR colorless to yellow and brown, massive varieties white or variously colored, red by hematite, etc.

LUSTER vitreous, usually pearly on cleavage planes.

PROPERTIES cleavage perfect parallel to the b plane, with a jagged fracture oblique to the a axis with some crackling noise, and with a distinct cracking sound parallel to c (see Fig. 21). Transparent to opaque.

CRYSTALS common, usually elongated, tabular, as shown in Figs. 21 and 28. Two types of twinning: swallowtail and Montmartre, the latter with curved faces. Acicular, roughly granular-cleavable also parallel-fibrous, compact as *alabaster*. Cleavage fragments and crystals of considerable size are not at all unusual, among them single crystals 50 cm. long and cleavage plates of even greater diameter. Thin cleavage flakes are fairly brittle, but somewhat thicker prisms may be bent quite easily. Curved crystals may occasionally also be found in nature.

IDENTIFICATION great softness (scratches readily with light

PLATE 39   *Pyroxenes*

Above left Gem-quality spodumene: pink kunzite from Pala, California, USA, and yellow orthoclase crystal from Itrongahy, Madagascar. *Center* Coarsely fibrous aggregate of diopside. Zillertal Alps, Tirol, Austria. *Center right* Perfect crystal of augite, as shown in Fig. 23. Firmerich, Eifel, Germany. *Below* Enstatite rock; crystal grains with bronze-colored shimmer on parting surfaces. Kraubath, Austria.

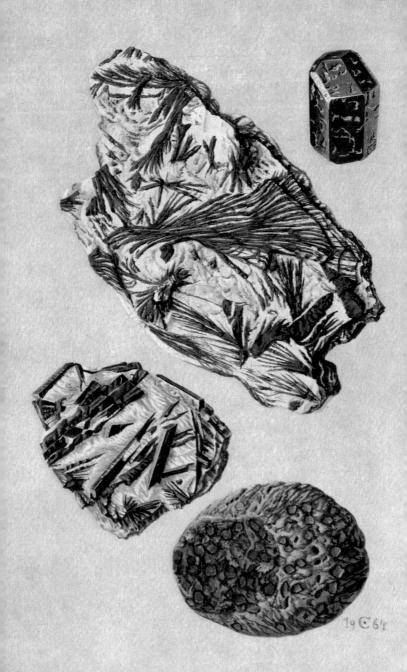

touch of finger nail!), crystal form, cleavage; no effervescence with HCl.

Gypsum is mentioned by Theophrasts as *gypsos*. Its use as burnt gypsum (for plaster), as well as alabaster was already known. Pliny in his *Historia Naturalis* mentions gypsum several times, although sometimes he does not distinguish between transparent gypsum and other minerals with good cleavages, such as calcite, barite, or mica; in particular, 'window glass' *(lapis specularis)* may mean cleavage flakes of gypsum as well as of mica. Among other things Pliny relates that beehives had been made from this material so that the bees could be observed at their work. Both ornaments and utensils made of compact alabaster have been in use since ancient times; Etruscan vases made of alabaster were found near Volterra, where alabaster is still quarried today.

Its use as a fertilizer dates from about the middle of the last century. Most gypsum is used in burnt form by the building industry. There are two types: in 'stucco gypsum', plaster of paris, part of the water is driven off by heating it to 125°C:

$$CaSO_4.2H_2O \rightarrow CaSO_4.\tfrac{1}{2}H_2O$$

The hemihydrate thus formed easily takes up water again and sets quickly. By heating (calcining) to 400°C, so-called flooring gypsum or anhydrite is formed:

$$CaSO_4.2H_2O \rightarrow CaSO_4$$

which takes up water more slowly but, on the other hand, becomes much harder. Quantities of gypsum are added to cement in order to control its setting time. In the chemical industry both gypsum and anhydrite are raw materials for the manufacture of ammonium sulphate fertilizer, sulphuric acid, and sulphur.

Formation and occurrence. Gypsum is almost exclusively formed

PLATE 40   *Amphiboles*
Above right Basaltic hornblende; crystal faces show etch marks. Bilina, Czechoslovakia. *Center* Hornblende schist; the crystals occur in thin, parallel, sheaf-like bundles. Zillertal Alps, Tirol, Austria. *Below left* Actinolite in talc schist. Zillertal Alps, Tirol, Austria. *Below right* Amphibolite, a rock consisting of amphibole needles and red garnet crystals. Partly polished boulder from the Isar river near Munich, Germany.

in sedimentary conditions, but not all by direct deposition. Much gypsum is derived from sedimentary anhydrite by subsequently taking up water; this is accompanied by a considerable increase in volume, and consequently, gypsum beds often appear to be swelled or contorted. Gypsum often occurs as a secondary formation in clays and marls, also in coal, frequently in very good crystals, and again in the weathering zone of sulphide ores. The required SO_4 is derived from decomposing pyrite. New crystallization of gypsum may often take place in salt-works from water containing gypsum. Sulphate-bearing solutions, which may occur in arid soils, are the cause of the formation of the so-called desert roses; these are platy crystals arranged as rosettes which include sand grains. Gypsum is occasionally deposited from fumaroles. It is usually accompanied by anhydrite, calcite, dolomite, limonite, and clay, in salt deposits sometimes also by boracite or quartz crystals, and not rarely by sulphur, which is often formed from gypsum. In gossans it is usually associated with limonite. Beautifully crystallized material may be found at innumerable places. Every salt deposit contains gypsum.

MEXICO – crystals (!!!), often huge (to 2 m.), in Mina Maravilla, Naica, Chihuahua, where the Cave of Swords within the mine preserves enormous bladed transparent crystals by the thousands. USA – clear 'textbook' crystals from shales and clays at Ellsworth, Mahoning Co.; bladed brown, sand-filled crystals (!!) in singles and clusters in briny floor of Great Salt Plain, Jet, Alfalfa Co., Oklahoma, some with 'hourglass' inclusions. ENGLAND – crystals (!!) at Trowbridge, Wiltshire and beautiful 'satin spar', strongly chatoyant, at East Bridgford, Nottinghamshire. GERMANY – crystals (!!) in copper shale at Eisleben, Thuringia; 'swallowtail' twins in clay at Schirnding, near Arzberg, Oberfranken, Bavaria; complete crystals in clay at Demmelswald, near Wiesloch, Baden. ITALY – sculpture *alabaster* at Volterra, Tuscany; crystals (!!) at the sulphur mines near Agrigento, Sicily. FRANCE – pale brownish 'swallowtail' twins, Montmartre, Paris.

SCHEELITE

CHEMISTRY – $CaWO_4$, calcium tungstate, with molybdenum commonly substituting for tungsten.

STREAK white   TETRAGONAL   H 4½–5   SG 6.1–5.9
COLOR colorless, white, yellowish, reddish, brownish, greenish or greyish.
LUSTER sub-adamantine.
PROPERTIES conchoidal to uneven fracture; brittle; transparent to opaque.
CRYSTALS fairly common as dipyramids resembling octahedrons but ranging from sharp to crude due to etching; also granular massive.
IDENTIFICATION luster, crystal shape, heaviness; fluoresces strongly under UV light!

Named after Swedish chemist K. W. Scheele; important ore of tungsten.

Formation and occurrence. In hydrothermal quartz veins, in pegmatites, and in contact metasomatic deposits. Commonly associated with cassiterite, wolframite, topaz, fluorite, tourmaline and mica, and with garnet, axinite, diopside in limestone-type contact deposits.

USA – large crystals (!!) to 15 cm. at Boriana mine, and Cohen mine, Cochise Co., Arizona; sharp grey octahedral crystals (!!) from Darwin, Inyo Co., California and from Greenhorn Mountains in the same state, as colorless transparent crystals (gem quality !!!). SWITZERLAND – small but very sharp crystals (!!!) Giuvstöckli. ENGLAND – sharp crystals (!!) Caldbeck Fells, Cumberland. GERMANY – crystals (!!) at Furstenburg, Saxony and Riesengrund, Riesengebirge. CZECHOSLOVAKIA – crystals (!!) at Cinvald (Zinnwald), Bohemia. ITALY – sharp crystals (!!!) at Traversella, Piedmont. KOREA – at Taehwa and other places, very large crystals (!!!) to several kg.

WULFENITE   Plate 29

CHEMISTRY $PbMoO_4$, lead molybdate, often with calcium substituting for lead.
STREAK white to yellowish   TETRAGONAL   H 3   SG 6.8
COLOR rarely colorless, usually yellow to orange, red, also grey or brown.
LUSTER adamantine.
PROPERTIES cleavage distinct, fracture conchoidal, brittle, transparent to translucent.

CRYSTALS common, as square plates, with modified edges, also pyramids and short prisms, thin platy-cellular, loosely porous, in crusts, rarely compact.

IDENTIFICATION fairly easy by color and habit, high luster, and by paragenesis.

Wulfenite was named after the Austrian mineralogist Father F. X. Wulfen, who in 1875 supplied a detailed description of this mineral with colored illustrations showing crystals. Wulfenite is an ore of molybdenum.

Formation and occurrence. Wulfenite is restricted to the oxidation zone of lead deposits, associated with weathered galena, cerussite, smithsonite, limonite, calcite, also with pyromorphite and vanadinite.

USA – Arizona is the prime state for magnificent wulfenite crystals, especially the Glove mine, Santa Cruz Co., where thin tabular crystals (!!!) were found measuring from 3–10 cm. on edge, also in splendid transparent red crystals (!!!) from Red Cloud mine and Hamburg mine, Yuma Co., and from Old Yuma mine, near Tucson, Pima Co. YUGOSLAVIA – beautiful crystals (!!) at Mežica, Slovenia. MEXICO – fine square tablets (!!!) from Villa Ahumada, Chihuahua. AUSTRIA – Bleiberg, Carinthia, provided the material described and depicted by Wulfen. SOUTH-WEST AFRICA – honey-colored thin tabular crystals (!!) from Tsumeb. CONGO (Brazzaville) – tan crystals (!!) to 10 cm. on edge from M'foati.

Class 7 Phosphates and vanadates

This section deals with compounds of metals with PO_4 (phosphate) or VO_4 (vanadate), with or without OH (hydroxyl) but other elements may also enter into the formula.

LAZULITE

$(Mg, Fe)Al_2(OH)_2(PO_4)_2$, a rare mineral, deep blue in color, named for its resemblance to lapis lazuli.

PLATE 41   *Micas*

Above Pseudo-hexagonal crystals of muscovite in calcite. Madagascar. *Below* Pink, massive lepidolite with smoky quartz and tourmaline. Mesa Grande, California, USA.

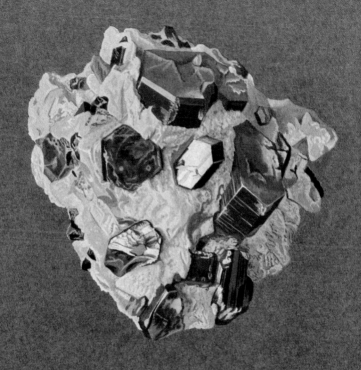

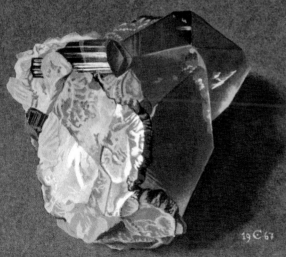

19 C 67

It occurs in small sharp crystals near Werfen, Salzburg, Austria and in fine large crystals (!!!) to 5 cm. in length, of doubly-terminated pyramidal habit, at the famous Graves Mountain locality in Lincoln Co., Georgia. Otherwise it occurs in granular masses in pegmatites and some metamorphic deposits (Champion mine, White Mountains, Inyo Co., California).

The apatite group   Plate 30

Apatite is undoubtedly the best-known mineral of the group. Pyromorphite and related minerals also belong here. All are hexagonal, and crystallize with similar habits, with partial substitution between end members. Cleavage is very indistinct. The composition of crystalline apatite is usually given as $Ca_5(PO_4)_3(F,Cl)$. Other apatites occur mostly in *phosphorite,* usually regarded as a rock and not a mineral.

		H	SG	Luster
Apatite series				
apatite	$Ca_5(PO_4)_3(F, Cl)$	5	3.2	vitreous–oily
hydroxyapatite	$Ca_5(PO_4)_3OH$			
oxyapatite	$Ca_{10}(PO_4)_6O$			
carbonate apatite	$Ca_{10}(PO_4)_6 \cdot (CO_3) \cdot (H_2O)$			
pyromorphite	$Pb_5(PO_4)_3Cl$	$3\frac{1}{2}-4$	7.0	adaman-tine
mimetite	$Pb_5(AsO_4)_3Cl$	$3\frac{1}{2}-4$	7.1	adaman-tine
vanadinite	$Pb_5(VO_4)_3Cl$	$2\frac{1}{2}-3$	7.0	adaman-tine

APATITE   Plate 30

Ca may also be replaced by Mg, Fe, Mn, Sr, Ba etc.

STREAK white

COLOR in all colors, particularly yellowish and greenish, blue and bluish green.

PLATE 42   *Serpentine minerals*
Above Chrysotile asbestos on serpentine rock. Black Lake, Quebec, Canada. *Below left* Yellowish-green gem serpentine. Spain. *Below right* Common serpentine. Wurlitz, Upper Franconia, Germany.

CRYSTALS common, in hexagonal plates, short and elongated prisms, slender prisms, even needles, and often with many faces. When massive, apatite is granular to compact, fibrous to coarsely radiating, or spathic as 'asparagus stone'. Phosphorite occurs as massive, porous aggregates, crusts, or as nodular, stalactitic, reniform and spherical concretions, earthy, oolitic or like chalcedony.

IDENTIFICATION crystal shapes, luster, hardness; easily distinguished from similar beryl and quartz crystals by hardness, from orthoclase grains by lack of cleavage. Crystals of nepheline are very similar but while apatite is completely and easily soluble in dilute acids, the silica in nepheline forms silica gel.

Apatite, named by Werner in 1789, is derived from the Greek *apatan,* to deceive, because it had been confused with beryl, quartz, and other hexagonal minerals. Apatite and phosphorite are the sole source of phosphates, 90 per cent of which is turned into fertilizers, the remainder into other chemicals. Phosphorus is also obtained as a by-product of iron smelting, since many iron ores contain some apatite.

Formation and occurrence. Abundant accessory mineral, and as such present in almost all igneous rocks. Now and again it may become enriched by differentiation, for example in magnetite-apatite ores or in the apatite deposits in nepheline syenite with nepheline, sphene, and aegirine, in the Khibin Hills, Kola Peninsula, USSR. It is widespread in all kinds of pegmatite, and may be pneumatolytic and hydrothermal. Apatite is typical of many alpine fissure assemblages. Lastly it is found in crystalline schists (in many talc schists as 'asparagus stone'). Phosphorite forms distinct marine sediments and also forms in massive deposits of bird guano, concentrated perhaps in pockets and cavities by the weathering of phosphatic limestones.

GERMANY – Greifenstein near Ehrenfriedersdorf, Erzgebirge, the type locality of apatite, in fine crystals (!!). Epprechtstein, Fichtelgebirge, Bavaria, fine crystals (!!) in pegmatite druses. AUSTRIA – Zillertal, Tirol, 'asparagus stone'; Radeckscharte and Grubenkarscharte, Hohe Tauern, crystals (!!!) in alpine fissures; Riffelkees, Stubachtal, with crystals of magnetite and olivine, diopside, tremolite, chlorite, copper ores, in matted chrysolite (mountain leather); Sölnkar, near Krimml, with epidote, and crystals of scheelite (!); Knappenwand (see epidote, p. 207) crystals (!!!). SWITZERLAND – at several places in the St Gotthard region, Graubünden, chief locality for fine, form-rich crystals (!!!) from alpine fissures as at Alpe de Sella, Piz Vallatscha etc., Zumdorf, Uri; near Hospental. ITALY – Biella, Piedmont, in syenite. Monte Somma, Vesuvius, crystals in druses.

CANADA – Renfrew Co., Ontario, crystals (!) in marble (up to 30 cm.). USA – Mount Apatite, Maine, transparent crystals (!!!, purple !). MEXICO – Cerro de Mercado, millions of fine transparent crystals (Pl. 30). BOLIVIA – Llallagua, crystals (!!!). PORTUGAL – splendid green crystals (!!!) with arsenopyrite, wolframite, etc. at Panasqueiras. BRAZIL – numerous pegmatites furnish clear to translucent crystals, often of gem grade. BURMA – blue, gem grade, rolled crystals in gem gravels.

For sharpness and perfection the crystals of Alpine vugs are difficult to surpass although recent finds at Panasqueiros, Portugal, and older specimens from Bolivia are also superb; the yellow crystals from Mexico are commonly cut into fine faceted gems to as much as 20 carats.

PYROMORPHITE   Plate 30

Occurs in oxidized portions of lead deposits as hexagonal prisms of generally small size (about $\frac{1}{2}$–1 cm.) but also elongated in subparallel growths (Bad Ems, Germany). Colors: brown, greenish, yellow, orange. GERMANY – fine brown crystals (!!!) Grube Rosenberg, Bad Ems, Rhineland; galena pseudos after thick tabular pyromorphite crystals from Bernkastel a.d. Mosel, Rhineland; Grube Clara, near Wolfach, Baden-Württemburg. USSR – large rude crystals (!!) Berezovsk, Urals. CZECHOSLOVAKIA – vivid green crystals (!!) Příbram. AUSTRALIA – brown crystals (!!!) at Broken Hill, New South Wales. USA – fine green crystals (!!!) in the Coeur d'Alene district, Shoshone Co., Idaho.

MIMETITE

Mimetite is rarer than pyromorphite and occurs in less fine crystals and commonly in globular aggregates. 'Campylite' variety, in strongly curved barrel-like crystals (!!!), occurs at Caldbeck Fells, Cumberland, England; bright orange globular in 79 mine, Gila Co., Arizona and Bilbao mine, Ojo Caliente, Zacatecas, Mexico; also fine from Tsumeb, South-West Africa.

VANADINITE

Occurs in oxidized zones of ore bodies containing lead, and found associated with pyromorphite, mimetite, wulfenite, descloizite, cerussite and limonite. Crystals hexagonal prisms, usually more perfect than pyromorphite; colors: red, brown, yellowish shades; translucent, rarely transparent; mostly as individuals of small size, about $\frac{1}{2}$–2 cm. USA – brilliant red crystals (!!) Apache mine, Gila Co.; similar crystals, but far less perfect, Old Yuma mine, near Tucson, Pima Co. MEXICO – the tapered crystals of brown 'endlichite' occur in fine specimens at Villa Ahumada, Chihuahua, also at Santa Eulalia. MOROCCO – beautiful brown-red tabular crystals (!!!) on rock, Mibladen; huge crystals (to 10 cm. long) at Djebel Mahseur.

VIVIANITE

$Fe_3(PO_4)_2 \cdot 8H_2O$, iron phosphate with water; white (colorless) when fresh but quickly turns blue when removed from the ground. Prismatic crystals, radiating fibrous, foliated, feathery aggregates, friable to earthy. Very fine crystals at Trepča, Yugoslavia, and Llallagua, Bolivia; giant crystals (up to $1\frac{1}{2}$ m.) found recently at Anloua near Ngaundere, Cameroon; fine crystals Bluebird district, Lemhi Co., Idaho and groups (!!!) near Richmond, Virginia, USA.

Identification: by color, gypsum-like cleavage and very low hardness.

TURQUOISE   Plate 30

CHEMISTRY $CuAl_6(PO_4)_4(OH)_8 \cdot 4H_2O$, copper aluminum phosphate with hydroxyl and water, some iron replaces aluminum.
STREAK white   TRICLINIC   H 5–6   SG 2.8
COLOR sky-blue, bluish-green, apple-green.
LUSTER waxy to dull.
PROPERTIES fracture conchoidal, fairly brittle, translucent to opaque.

CRYSTALS very rare, minute; most gem material is compact, nodular ('nuggets') or in seams or veinlets.
IDENTIFICATION color, massive forms.

The name 'turquoise' – the French word for 'Turkish' – probably dates from the sixteenth century, when the mineral, found in Persia, was brought from there to Europe via Turkey. It was in use as an ornamental stone in Neolithic times, and on the Sinai peninsula it was quarried six thousand years ago. Aristotle, in his *De Petris,* says, 'Its color cheers the sorrowful.' In the New World, the Aztecs knew turquoise as *chalchihuitl* (though in some places this name was also given to jade). Stones with fine limonite or manganese oxide veins are called 'matrix-turquoise'. Guaranteed color-fastness is highly prized, since many turquoises tend to fade or change their color, owing to the action of soap, cosmetic creams, perfumes and other chemicals. Impregnating the stones with synthetic resins seems to increase their resistance. Substitutes are widely used.

Formation and occurrence. Turquoise veins are formed as a decomposition product of Al-rich rocks, provided that Cu-sulphides are present. Only a small fraction of the available material is suitable for polishing. EGYPT – Wadi Meghara and other places in the Sinai peninsula; ancient localities. IRAN – Maden, Nishapur district, Khorassan, for over three thousand years the source of particularly beautiful 'oriental' turquoise. USA – numerous mines in Nevada, a record 'nugget' of 70 kg. came from near Battle Mountain, Lander Co., and another of 80 kg. from Lynn district; numerous mines also in Arizona, where good turquoise sometimes occurs in the open-pit copper mines, and in New Mexico; druses of microcrystals (!) occur on gneiss near Lynch Station, Campbell Co., Virginia. CHILE – some fair turquoise has been found in copper mines of Chuqui-camata.

Class 8 The silicates

The SiO_4 tetrahedron is the basic unit common to all silicates. Tetravalent silicon is always surrounded by four oxygen atoms situated at the corners of a tetrahedron (Fig. 22). In the simplest case isolated SiO_4^- groups are present in the structure, in minerals like olivine (Mg, Fe)$_2$SiO$_4$. Silicate tetrahedrons may, however, also be joined at their corners. Two such tetrahedrons may share an oxygen to give Si$_2$O$_7$ groups, while 3, 4, or 6 tetrahedrons may share two oxygens each to form rings; silicate structures of this type are termed *finite,* with individual silicate units joined to one another by metal atoms. SiO$_4$ tetrahedrons may also be joined together forming *infinite* chains (Fig. 22). The structures of pyroxenes contain such single chains whereas in ampiboles two chains are joined side-by-side, forming a double chain. The minerals in this group tend to have elongated, prismatic habits or even be fibrous, with a good cleavage parallel to their elongation.

A third possibility arises if the tetrahedrons are joined together forming *sheets,* so that they are now infinite in two directions. These sheets by themselves have relatively little electric charge for their size, so that neighbouring sheets are not held together very strongly. As may be expected, minerals belonging to this group are platy or flaky, with a perfect cleavage parallel to the plates as in the micas. In the *framework* silicates the tetrahedrons are joined at all four corners. Quartz, SiO$_2$, really belongs here but is traditionally included among the oxides. In such a framework a complete balance of valencies is achieved. Other compounds are possible only because trivalent aluminum may replace tetravalent silicon. Potassium

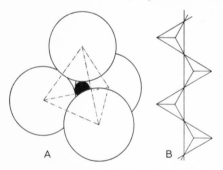

22   Silicate structures. (A) A silicate tetrahedron (black sphere silicon, white spheres oxygen), the unit of which all silicates and minerals of the quartz group are built. (B) Chain of SiO$_4$ tetrahedra (in plan view, without showing the atoms), each joined to the next by a shared oxygen atom. Double chains, rings, sheets etc. can be built in the same manner.

feldspar K^+ $(AlSi_3O_8)^-$ is just one example. The silicates may thus be classed into four groups:

1 Isolated groups: single or double tetrahedrons (SiO_4 or Si_2O_7 groups) and ring-silicates (3-, 4-, or 6-membered rings as $Si_6O_{18}^{12-}$)
2 Single and double chains $(Si_2O_6)^{4-}$ and $(Si_4O_{11})^{6-}$
3 Sheets $(Si_4O_{10})^{4-}$
4 Frameworks $(Si_4O_8)^0$; valency depends on extent of Al substitution.

About a third of all minerals are silicates, including the major rock-forming minerals.

Section 1 Isolated tetrahedrons and ring silicates

OLIVINE SERIES Plate 33

CHEMISTRY (Mg, Fe)$_2$SiO$_4$, magnesium-iron silicates; a complete isomorphous series extends between the end-members *forsterite* (Mg$_2$SiO$_4$) and *fayalite* (Fe$_2$O$_4$). Alters readily to serpentine.

STREAK colorless   ORTHORHOMBIC   H 6½   SG 3–4

COLOR forsterite white, assuming increasingly greenish-yellowish shades with increasing iron content, with fayalite being black.

LUSTER vitreous-oily.

PROPERTIES poor cleavage, conchoidal fracture; brittle; transparent to opaque.

CRYSTALS uncommon; boxlike, stubby to moderately long prismatic, often with striated sides; St John's Island crystals with knifelike edges; commonly fine to coarse granular (Pl. 33).

IDENTIFICATION characteristic green hue in basaltic rocks; association with serpentine, black spinel, chromite; softness (softer than green tourmaline).

Olivine, named after its color, is also used as a gemstone when it occurs in the transparent green variety known as *peridot* (older name, 'chrysolite'); it is probably the 'green topaz' mentioned by Pliny and was much treasured and used in medieval times to decorate church plate and robes. Fayalite is uncommon except in the slags which appear during smelting of iron; it was named after Fayal Island, Azores, where it occurs in volcanic rocks.

Formation and occurrence. Most olivine appears as grains and aggregates ('peridot bombs') in basalts and also gabbros and norites, while dunite and peridotite are rocks composed primarily of olivine; kimberlites contain relicts of olivine with most altered into serpentine; large intrusive igneous rock masses of peridotite are commonly wholly or partly altered to serpentine. Forsterite occurs in dolomite rocks and contact-metamorphic limestones such as the ejecta of Monte Somma, Vesuvius. Also occurs in Fe-Ni meteorites.

EGYPT – Zebirget (St John's Island), Red Sea, is the classic source of fine large crystals (!!!) used for gems. BURMA – gem crystals (!!!) from metamorphosed limestones, commonly to 8 cm. long. CEYLON – grains in gem gravels. NORWAY – unique pseudomorphs of serpentine after olivine at Snarum. USA – large quantities of small gems are cut from the granular peridot 'bombs' in basalt on the San Carlos Indian Reservation, Gila Co., Arizona; crystals (!) of forsterite at Bolton, Massachusetts.

The garnet group   Plates 31, 32, and 40

The garnets comprise a number of minerals with similar crystal structures, external crystal forms, and general appearance despite their varying chemical compositions. The general formula is usually written as a metal + aluminum silicate: $(M^{2+})_3(M^{3+})_2(SiO_4)_3$, where magnesium (Mg), iron (Fe), manganese (Mn), calcium (Ca) occupy the divalent (M^{2+}) positions and aluminum (Al), but also Fe^{3+} and Cr^{3+}, occupy the trivalent (M^{3+}) positions. Traces of other elements are commonly present. Few, if any, garnets are chemically 'pure', most showing mixtures of ions. Thus most iron garnet *almandite,* contains considerable manganese, and indeed a complete series toward the manganese end-member *spessartine* exists; other, though not necessarily complete, substitutions occur between the other end-members shown below with their idealized formulas.

PYROPE $Mg_3Al_2(SiO_4)_3$ – the magnesium garnet; blood-red, often very dark; 'Bohemian garnet', 'Cape ruby', etc.

ALMANDITE $Fe_3Al_2(SiO_4)_3$ – the iron garnet; dark brown-ish-red, red, purplish-red, to nearly black (a purplish variety known as 'rhodolite' contains much of the pyrope molecule).

SPESSARTINE $Mn_3Al_2(SiO_4)_3$ – the manganese garnet; yellowish to orange to red-orange.

GROSSULAR $Ca_3Al_2(SiO_4)_3$ – the calcium garnet; colorless, white, greenish, yellowish, orange, red (an orange variety is known in the gem trade as 'essonite' or 'hessonite').

ANDRADITE $Ca_3Fe_2(SiO_4)_3$ – calcium-iron garnet; green, brown, yellowish, black ('melanite'); also vivid emerald-green ('demantoid').

UVAROVITE $Ca_3Cr_2(SiO_4)_3$ – calcium-chromium garnet; rare; emerald-green to dark green.

STREAK colorless   CUBIC   H $6\frac{1}{2}$–7   SG 3.5–4.2
COLOR as above but never naturally blue or purple.
LUSTER vitreous, sometimes somewhat resinous.
PROPERTIES absence of cleavage; fracture conchoidal to uneven, commonly garnets break up into small box-like fragments; brittle; transparent to nearly opaque.
CRYSTALS very common as rhombic dodecahedrons, trapezo-hedrons, and complex forms modifying both; usually ball-like in general shape; from minute to some 60 cm. diameter (almandite, New York), also granular massive.
IDENTIFICATION best clue is crystal shape; also relative heavi-ness, and colors; hardness.

The name garnet was originally given because of its similarity to seeds of the pomegranate. 'Pyropos' means 'fiery-eyed'. 'Almandite' is very likely a corruption of Alabanda, an ancient city in SW. Anatolia where, according to Pliny, cutting and polishing of garnets was practised. Spessartine was originally found in Spessart, Germany. 'Grossular' is derived from the Latin *grossularia,* gooseberry. Andradite was named after the Portuguese mineralogist d'Andrada, and uvarovite after the Russian statesman Count Sergei Uvarov. The color of topaz-olite is reminiscent of topaz, and demantoid is named after its adamantine luster. Theophrastus describes red gemstones as *anthrax,* Pliny as *carbunulus,* both names for incandescent coals.

Ruby, red spinel, and other red stones were, no doubt, also masquerading under the two names pyrope and almandite.

The extensive use of red garnet in ancient times has been established from the many finds of jewelry, ornaments and utensils. Pyrope, particularly from Bohemia, was a fashionable stone of the last century. Pyrope from South Africa is known as 'Cape Ruby'. In places like the Austrian Tirol almandite polishing and cutting is still carried on. Garnet is a long-lasting abrasive especially suitable for wood, glass and plastics. Some time ago there were several garnet mills in the Alps, but none of these is now in operation. On the other hand, about 10,000 tons are produced yearly in the United States (Gore Mountain, see below).

Formation and occurrence. Garnets occur mainly in metamorphic rocks: almandite particularly in mica schists, gneisses, amphibolites, eclogites (Ca-Mg-bearing almandite with augite); grossular and andradite in contact-metamorphic limestones, skarns, and their special mineral assemblages ('calcium-silicate rocks' with idocrase, diopside, epidote, hedenbergite, hornblende, wollastonite, chlorite, magnetite and sulphides, Fe-bearing sphalerite, for example, also with scheelite). Pyrope occurs in basic rocks, such as kimberlite and in serpentines derived from them. Spessartine is a pegmatite mineral and almandite too is partly found in pegmatites. Uvarovite always occurs with chromite. Demantoid is found only in serpentinites. Melanite is a rare minor constituent of volcanic rocks, such as phonolite. Garnets are quite commonly found in gravels as well as in sands.

AUSTRIA – almandite in mica- and chlorite-schists at Granatkogel, near Obergurgl in the Ötztal Alps, Tirol; Pusygraben near Lölling, Carinthia, up to 12 cm. in diameter; St Leonhard, Saualpe, Carinthia, large crystals in mica-pegmatite; Längenfeld near Ötz, Tirol, eclogite and garnet amphibolite. SWITZERLAND – near Zermatt and Geisspfad, Binnatal, Valais, andradite (demantoid) along the marble-serpentine contact; Val Tremola, St Gotthard massif, Graubünden, particularly large almandite crystals; Val Maighels, northern Gotthard massif, hessonite; near Disentis, Graubünden, topazolite with zeolites; Claro, Val de Molino, Ticino, calcium-silicate rock with hessonite, epidote, idocrase. ITALY – Ala valley, Alto Adige, hessonite

(!!!) with diopside, idocrase, etc. in serpentinites; Traversella, Piedmont, pyrite, magnetite, garnet, scheelite. GERMANY – Laacher See, Eifel; Auerbach, Hesse, grossular (!!) in contact marble with idocrase, wollastonite, epidote; in the vicinity of Aschaffenburg, Bavaria, spessartine in pegmatite; Wurlitz, Fichtelgebirge, topazolite in serpentine. FINLAND – Outokumpu, N. Karelia, Cr deposit with uvarovite (!!!). CZECHO-SLOVAKIA – at Bilina and elsewhere in Bohemia, classic pyrope localities. USSR – classic demantoid locality Sissersk district, Nizhni-Tagil, Urals. USA – huge almandite crystals to as much as 1 m. diameter at Gore Mt, Warren Co., New York, sometimes furnishing clear gem material; andradite ('polyadelphite') in crystals to 10 cm. from the zinc mines of Franklin-Ogdensburg, Sussex Co., New Jersey; trapezohedrons of andradite at French Creek mines, Chester Co., Pennsylvania; gem 'rhodolite' in gravels of Cowee Creek, Macon Co., North Carolina; beautifully formed crystals (!!!) of almandite along Skeena and Stikine rivers, British Columbia (Canada) and Wrangell, Alaska; large simple dodecahedrons of almandite (!!) in Sedalia mine, Chaffee Co., Colorado, to 25 cm. diameter; 'star' almandite garnet, much used for gems, from gravels of Emerald Creek, Benewah Co., Idaho; andradite crystals (!!!) with quartz and epidote at Garnet Hill, Calaveras Co., California; beautiful spessartine crystals (!!!) on albite with black tourmaline in pegmatites of Ramona district, San Diego Co., California; gem pyrope (!!!) from deposits near junction of Utah, Arizona and New Mexico. CANADA – magnificent transparent, pale orange grossular crystals (!!!) many useful for gems, in seams in serpentine rocks near Asbestos, Quebec. MEXICO – enormous numbers of white, pink and greenish grossular crystals (!) of simple dodecahedral form are furnished from marble around Lake Jaco, Chihuahua.

ZIRCON   Plate 35

CHEMISTRY $ZrSiO_4$, zirconium silicate, with Fe, Ca, U, Th.
STREAK white   TETRAGONAL   H $7\frac{1}{2}$ SG 4.5
COLOR brownish-red, brown, yellow, grey, green, also colorless, blue by heating.
LUSTER adamantine or resinous.

PROPERTIES cleavage indistinct, conchoidal to uneven fracture, brittle; transparent to nearly opaque.

CRYSTALS common, simple combinations of tetragonal prisms and pyramids, otherwise in grains, never in massive aggregates.

IDENTIFICATION crystal form and typical luster.

The name is supposed to have been derived from the Persian words *zar,* gold, and *gun,* color. The brown and yellowish-red crystals are known as 'hyacinth' in the trade. Metallic zircon is of importance in the construction of nuclear reactors and for special alloys. Zirconium oxide is used for making high-temperature resistant crucibles, enamels and glazes, etc., and chief sources are from beach sands of Australia, Brazil, India etc.

Formation and occurrence. Zircon is a minor accessory mineral primarily of acid igneous rocks and also of crystalline schists; larger crystals occur in certain pegmatites. Zircon is chemically resistant and hence commonly appears in gravels. Gem-quality zircons come from the gem gravels of Ceylon, Burma and Thailand. CANADA – brown crystals (!), sometimes to nearly 7 kg. weight, found in various syenite rock occurrences in Bancroft and Hastings counties, Ontario, also fine 'cyrtolite' (radioactively altered variety) in Dungannon township, Hastings Co. U.S.A – small sharp brown crystals (!) on St Peter's Dome, El Paso Co., Colorado. NORWAY – fine crystals (!!) in nepheline-syenite pegmatites at Kragerö and elsewhere in Southern Norway. USSR – sharp crystals (!!) near Miass, Ilmen Mountains. CEYLON (and BURMA) – gem gravels, sharp to rounded prisms (!–!!!). AUSTRALIA – fine lustrous crystals (!!) Harts Range, Northern territories; also in gravels in New England district, New South Wales.

SILLIMANITE, ANDALUSITE, KYANITE

These three minerals have the same chemical composition, Al_2SiO_5. They occur in different crystal forms and have widely different external, but the same chemical, characteristics. All three are typical accessory minerals of crystalline metamorphic rocks.

All three are metamorphic minerals found especially in gneisses, mica schists, and some eclogites. Quite sizeable crystals may occur in fissures and quartz and pegmatite veins. Andalusite is also associated with contact-metamorphic, aluminous rocks. Andalusite and kyanite can often occur together with staurolite, while sillimanite is usually accompanied by corundum, spinel, cordierite and andalusite. Muscovite and quartz are nearly always also present. Compact aggregates of kyanite or sillimanite are important raw materials. At a temperature of about 1550°–1600°C these minerals alter to a finely matted aggregate of the fibrous mineral mullite with silica glass. This material is used for making porcelain-like utensils which are extremely resistant to heat and the action of chemicals.

Important deposits occur in India (Bihar, kyanite and sillimanite; Assam, kyanite, andalusite, corundum), USA (North Carolina, California), Russia, Kenya and South Africa.

SILLIMANITE

Named after U.S. mineralogist B. Silliman. Usually fibrous aggregates of silky luster, and very rarely in crystals although pale blue gem-quality crystals do occur in the gem gravels of Mogok district, Burma and have been cut into gems. Some fibrous compact material from Idaho, USA has been used for chatoyant gemstones.

ANDALUSITE   Plate 33

STREAK white   ORTHORHOMBIC   H 7½   SG 3.1
COLOR in various colors, rarely also colorless and clear.
LUSTER vitreous.
PROPERTIES cleavage indistinct, fracture uneven, transparent to opaque; often superficially altered to muscovite (sericite).
CRYSTALS not uncommon, usually prismatic with nearly square cross-section; granular. *Chiastolite* contains clayey or carbonaceous inclusions embedded in such a manner that cross-shaped figures appear in slices taken through the crystals (Pl. 33).
IDENTIFICATION crystal form; coating of mica, strong color change in transparent gem crystals; cross figures in chiastolite.

Andalusite was named after occurrence in Andalusia, Spain. *Chiastolite* refers only to material with a cross-shaped inclusion. Chiastolites were once used as amulets.

Formation and occurrence. In gneisses, schists as a metamorphic mineral. Chiastolite is abundant in many places in the world and only a few occurrences can be mentioned; on the other hand, transparent crystals come only from Brazil, and translucent to opaque prisms of non-gem quality appear commonly in pegmatites, and some metamorphic occurrences. BRAZIL – crystals of poor to good form, gem quality, olive-green to brownish-pink, and displaying strong color change (trichroism) in gravels near Santa Tereza, Espirito Santo, and Arassuahy, Minas Gerais; cuttable to 15 carats flawless. AUSTRIA – famous occurrence of large squarish prismatic crystals (!!!) at Lisenz Alpe, Tirol. CEYLON (and BURMA) – good gem crystals but water-worn, in gem gravels. USA – fine chiastolite at Lancaster, Worcester Co., Massachusetts, many localities in Kern, Mariposa, and Madera counties, California. AUSTRALIA – fine chiastolite at Bimbowrie, South Australia.

KYANITE   Plate 33

STREAK white   TRICLINIC   H 6–7 at right angles to elongation, parallel to it 4   SG 3.6

COLOR colorless, white, frequently blue, green, also colored in zones.

LUSTER vitreous to pearly on cleavage planes.

PROPERTIES cleavage, parallel to large face, perfect; fibrous fracture; transparent to translucent.

CRYSTALS common, bladed, often undulating curved, horizontally striated, foliated, radiating; tangled fibres.

IDENTIFICATION shape of crystals; dual hardness; when finely fibrous, difficult to distinguish from andalusite or sillimanite.

Kyanite gets its name from its color (Greek *kyanos,* blue). It is also known as *disthene,* from the Greek *dis,* dual, and *sthenos,* strength – alluding to its two widely differing hardnesses.

Occurrence. BRAZIL – Capelinha, Minas Gerais, elongated platy crystals up to 12 cm. long, sometimes of gem quality. SWITZERLAND – Alp Arena, Campolungo, Passo Cristallina, Pizzo Forno (Pl. 33) and elsewhere in Ticino; the classic locality on Pizzo Forno (kyanite and staurolite crystals in paragonite) is often wrongly given as 'Monte Campione'. AUSTRIA – Greiner, Zillertal Alps, Tirol; Millstädter Alps and

Laufenberg near Radenthein, Carinthia. TANZANIA – large (to 30 cm.) crystals, blades (!) of blue color, sometimes cuttable into faceted gems, from near Sultan Hamud. CEYLON – clear facet-grade crystal fragments in gem gravels. USA – fine green crystals (!!) in Yancey Co., and elsewhere in North Carolina.

TOPAZ   Plate 34

CHEMISTRY $Al_2SiO_4(F, OH)$, aluminum silicate with fluorine and hydroxyl, usually very pure.

STREAK colorless   ORTHORHOMBIC   H 8   SG 3.5

COLOR primarily colorless, whitish, also faint blue to pale blue, seldom dark; more rarely, yellow, orange, purplish, reddish, brownish, greenish.

LUSTER vitreous.

PROPERTIES perfect and easily developed cleavage along basal plane, seldom absent; also conchoidal to irregular fracture; brittle; transparent to translucent.

CRYSTALS common as pointed wedge-like individuals with diamond-shaped cross-sections; also squarish, blocky; seldom elongated; rarely fine granular to coarse granular massive ('pycnite').

IDENTIFICATION perfect cleavage plus crystal shape; hardness; associations with pegmatite minerals.

Cut topazes from ancient times are known. The name is derived partly from a legendary island, Topasos, in the Red Sea, and partly from *tapas,* the Sanskrit word for fire. Yellow, orange, and reddish topazes are popular gemstones. Colorless topazes (known in Brazil as *Pingas d'agua,* waterdrop) were used as imitation diamonds; the 'Braganza' stone of the Portuguese crown (over 1,600 carats) was found to be a topaz after it had, for a long time, been regarded as an extremely valuable diamond. The Royal Ontario Museum in Toronto and the Smithsonian in Washington, D.C. possess blue cut gems of over 3,000 carats while the Field Museum in Chicago owns an even larger example. Single crystals of topaz often reach enormous sizes, well in excess of 50 kg.!

Formation and occurrence. Topaz is primarily a granitic pegmatite species of hydrothermal origin, but also occurs as a pneumatolytic mineral in cassiterite veins with fluorite, wolframite, tourmaline, quartz, beryl, and micas. The finest crystals are obtained from vugs in pegmatites. Good physical and chemical resistance allow topaz to concentrate in gravels, sometimes in quantities mineable for gem material. BRAZIL – numerous localities (pegmatite) in Minas Gerais, yielding fine crystals (!–!!!) to many kilos' weight, primarily colorless or pale blue; the finest gem topazes (yellow, orange, pink, red-purple) come from a quartz vein system near Ouro Preto which has furnished rough since the seventeenth century. USSR – superb sharp blue crystals (!!!) from pegmatites in vicinity of Mursinka, Ala-bashka, etc., NE. of Sverdlovsk, and in the Ilmen Mountains, Urals; the crystals are implanted on matrix and are generally quite clear; also in large crystals of fair quality from pegmatites in the Ukraine. USA – fine blue crystals (!!), resembling some-what the Uralian, at Little Three mine, Ramona, California; small sharp colorless to sherry color from vugs in volcanic rock Thomas Mountains, Utah; in large sherry color crystals from pegmatites in the Pikes Peak granite of the Front Range, Colorado (sherry color fades in sunlight or by heating!). MEXICO – single, sharp and lustrous crystals (!!) in volcanic rocks from a number of places in San Luis Potosi. SOUTH-WEST AFRICA – fine crystals (!!) in pegmatites around Kleine Spits-kopje. JAPAN – crystals (!!) in vugs in pegmatites in Omi and Mino provinces. GERMANY – classical locality for small pale-yellow crystals in vugs of the Schneckenstein, Saxony; fine blue crystals (!!) in pegmatite vugs of the Fichtelgebirge.

STAUROLITE   Plate 33

CHEMISTRY $Fe_2Al_9O_7(OH)(SiO_4)_4$, iron aluminum hydroxyl silicate.

STREAK white   ORTHORHOMBIC   H $7–7\frac{1}{2}$ SG 3.7
COLOR reddish-brown to blackish-brown.
LUSTER vitreous.
PROPERTIES cleavage distinct in one direction, fracture uneven to splintery; opaque, translucent, rarely transparent.

PLATE 43   *Analcime, leucite, lapis lazuli (lazurite)*
Above A specimen with large trapezohedrons of analcime. Italian Alps. *Center right* Lapis lazuli in the form of a crystal (a rarity). Badakhshan, Afghanistan. *Below* One large, well-developed crystal of leucite and several smaller ones, in lava. Vesuvius, Italy.

19C67

CRYSTALS common as prismatic crystals and as interpenetrant twins.

IDENTIFICATION twinned crystals, usually with adhering mica.

Stauros is the Greek for 'cross'. The twinned crystals have for a long time been known as 'cross stones' and Catholics have worn them as amulets. In the Southern States of the USA they are still sold as lucky stones today.

Formation and occurrence. Staurolite is formed very similarly to andalusite and kyanite, in metamorphic rocks, particularly gneisses and schists. USA – Patrick Co., Virginia; Fannin Co., Georgia, and Pilar, Rio Ariba Co., New Mexico, good crystals and twins (!!). AUSTRIA and ITALY – in several mica schist outcrops, as the Southern Ötztal Alps, in the Radenthein area of Carinthia, and at Vipiteno (Sterzing) just across the Brenner Pass in Italy. SWITZERLAND – in Ticino with kyanite (!!). USSR – Sanarka River, S. Urals. GERMANY – Kleinostheim, NW. of Aschaffenburg, Bavaria, crystals in staurolite gneiss.

SPHENE (titanite) Plate 35

CHEMISTRY $CaTiSiO_5$, calcium titano-silicate.

STREAK white   MONOCLINIC   H 5–5$\frac{1}{2}$ SG 3.5

COLOR yellow, brown, green, grey, reddish-brown, rarely black.

LUSTER adamantine.

PROPERTIES cleavage indistinct, fracture conchoidal, brittle, transparent to opaque.

CRYSTALS common, either as flat, envelope-shaped crystals, wedge-shaped in cross-section, or more tabular, also short prismatic very commonly as interpenetrant twins (Pl. 35); occasionally massive granular.

IDENTIFICATION crystal forms, high luster, twinning.

Common sphene, usually called titanite, is a constituent of plutonic rocks, while the name sphene is reserved for well-

PLATE 44   *Feldspar: 1 – potassium feldspars*
Above left Orthoclase crystal as shown in Fig. 24B, p. 244, with additional modification of edges. Ochsenkopf, Upper Franconia, Germany. *Center* Form-rich crystal of microcline, 'amazonite'. Pike's Peak region, Colorado, USA. *Below left* Carlsbad twin, as in Fig. 24D, p. 244. Weissenstein, Upper Franconia, Germany.

developed or twinned crystals. Clear examples are cut into gems.

Formation and occurrence. Sphene is a minor constituent of many plutonic rocks and of some crystalline schists. Sphene occasionally occurs in contact-metamorphic limestones together with diopside, garnet, epidote and others, and in pegmatites (particularly in syenite pegmatites). The best sphene is found in alpine fissures, with calcite, albite, adularia, chlorite.

SWITZERLAND – crystals (!–!!!) widely found in alpine fissures in the Aar and Gotthard massifs, and at some localities in the Pennines; crystals commonly green, yellow, brownish, and twinned, sometimes to 5 cm. long. AUSTRIA – fine crystals (!–!!!) similar in occurrence to Swiss in Tirol and Salzburg, with superb crystals once found in the Untersulzbachtal and Felbertal, Salzburg. CANADA – dark brown 'envelope'-shape crystals in Grenville marbles, notably at Eganville, Renfrew Co., Ontario (crystals to 20 cm. diameter). USA – dark brown crystals in Grenville marbles at various places in St Lawrence Co., New York; green, clear crystals to 4 cm. diameter (gem quality in part) from Tilly Foster iron mine, Brewster, Putnam Co., New York. MEXICO – in small pegmatites as tabular crystals of yellow, green to brown hues, some to 10 cm. diameter and partly of gem quality, in a large area of Baja California south of Ojos Negros to San Quintin; at the latter Cr-bearing sphene of beautiful emerald green color has been found.

EPIDOTE and CLINOZOISITE   Plate 32

CHEMISTRY $Ca_2(Fe, Al)Al_2SiO_4.Si_2O_7(O, OH)$, calcium iron aluminum silicates; *piemontite* has a noticeable Mn content, replacing Fe; clinozoisite is iron-deficient epidote.

STREAK white   MONOCLINIC   H 6–7   SG 3.4

COLOR crystals varying shades of dark green to blackish, more rarely yellowish-green, even grey, massive varieties generally paler green. Piemontite is cherry-red to reddish-black, clinozoisite greyish, greenish, brownish.

LUSTER vitreous.

PROPERTIES one good cleavage direction, fracture uneven; brittle; translucent to transparent.

CRYSTALS nearly always elongated parallel to the *b*-axis, very form-rich, elongated faces striated, twinning common. Radiating to compact; piemontite and clinozoisite nearly always occur as tiny needles, though piemontite can also have radiating or granular appearance.

IDENTIFICATION color and crystal form; the color of epidote is a typical 'pistachio green'.

The word epidote is derived from the Greek *epidosis;* 'pistacite' (massive epidote) gets its name from its pistachio-green color. Piemontite is named after its occurrence in a manganese deposit near San Marcello, Val d'Aosta in Piedmont, Italy. The name clinozoisite, 'inclined zoisite', indicates the difference from zoisite, the orthorhombic mineral. As 'green topaz' (schorl) from Dauphiné, epidote was described as early as the eighteenth century.

Formation and occurrence. Epidote is an important constituent of metamorphic rocks, although it is invariably formed only in the presence of hot solutions. Together with idocrase, grossular etc., it is typical of contact-metamorphic limestone and skarn assemblages. The most beautiful crystals come from the specialized assemblages associated with alpine fissures. ITALY – Val di Fassa, Dolomites (Alto Adige); Mussa, Ala valley, Piedmont. AUSTRIA – Knappenwand, Untersulzbachtal, Hohe Tauern, the best-known epidote locality, with crystals (!!!) up to 20 cm. long, associated with calcite crystals, very thin actinolite needles (byssolite), apatite, albite, adularia, crystals of scheelite. SWITZERLAND – Kammegg near Guttanen, Bern, epidote with amphibole asbestos (amiant), well crystallized scheelite, adularia, chlorite.

USA – superb crystals and groups (!–!!!), with quartz (Japanese twins!) and adularia (twins!) at Green Monster Mountain, Prince of Wales Island, South-eastern Alaska, where groups of shining perfect crystals as much as 30 cm. in diameter have been recovered from cavities in skarn rocks, with individual crystals to 10 cm. long; lustrous stubby crystals in Calumet iron mine, Chaffee Co., Colorado; large crystals in Greenhorn Mountains, Kern Co., California. MEXICO – clinozoisite and epidote crystals in scheelite skarn deposits, Los Gavilanes district, Baja California, also with sphene in rude crystals in the Ojos Negros–San Quintin district and in cavities in volcanic rocks along seashore south of Ensenada. FRANCE – fine crystals. (!!) Auris, near Bourg d'Oisans, Dauphiné. NORWAY – crystals (!) Arendal. MOZAMBIQUE – recently as very simple, large (to 10 cm.) crystals.

ZOISITE

With same chemical formula as clinozoisite but orthorhombic in crystallization; a common rock-forming mineral; a fibrous pink variety, sometimes compact enough to be cut into cabochon gems, is known as *thulite,* with the best specimens coming from Norway.

HEMIMORPHITE

CHEMISTRY $Zn_4(OH)_2(Si_2O_7).H_2O$, zinc hydroxyl silicate with water.

STREAK colorless   ORTHORHOMBIC   H $4\frac{1}{2}$ SG 3.4

COLOR usually white or colorless; sometimes colored by inclusions.

LUSTER vitreous.

PROPERTIES perfect cleavage; uneven to conchoidal fracture; brittle; translucent to transparent.

CRYSTALS thin tabular to bladed, usually small; one end of crystal with blunt termination, other with sharp termination (hemimorphic).

IDENTIFICATION white bladed crystal groups in fans, sprays, etc.; soluble in hydrochloric acid, forming jelly.

Named after 'hemi' and 'morph', literally 'half-form', in allusion to crystal forms. A minor ore of zinc.

Formation and occurrence. In oxidized zones of lead-zinc deposits. MEXICO – beautiful sprays, sometimes with red (hematite?) inclusions from Mina Ojuela, Mapimi, Durango, and in large reddish crystals to 5 cm., from Santa Eulalia, Chihuahua. USA – crusts of 'maggot ore', so called because of the curious worm-like upper surfaces of the masses, white, from Ogdensburg, Sussex Co., New Jersey. USSR – crystals

PLATE 45   *Feldspar: 2 – potassium feldspars*
Above left Simple crystal of adularia with chlorite. Switzerland. *Above right* Adularia, with some bluish sheen: 'moonstone'. Zillertal Alps, Tirol, Austria. *Below* Platy sanidine in trachyte. Drachenfels, Germany.

(!!) Nerchinsk, Transbaikalia. GERMANY – Altenberg, Saxony. ITALY – crystals (!!) at Iglesias, Sardinia. ALGERIA – crystals (!!!) at Djebel Guergour, Constantine.

ILVAITE

A calcium-iron, hydroxyl silicate, $CaFe_2O(OH)Si_2O_7$, found as black orthorhombic crystals of diamond-shaped cross-section in some iron deposits as at Laxey mine, South Mountain, Owhyee Co., Idaho and from Rio Marina and Capo Calamita, Elba, Italy; also crystals from Seriphos Island, Cyclades, Greece, and from Japan.

IDOCRASE (VESUVIANITE)   Plate 36

CHEMISTRY complex Ca-Mg-Fe silicate of still somewhat uncertain formula, $Ca_{10}Mg_2Al_4(OH)_4(Si_2O_7)_2(SiO_4)_5$.
STREAK white   TETRAGONAL   H $6\frac{1}{2}$ SG 3.4
COLOR brown or green; also grey, yellow, reddish-brown to
black, rarely blue, red or pink.
LUSTER vitreous.
PROPERTIES cleavage indistinct, fracture uneven, translucent,
rarely transparent.
CRYSTALS common, usually prismatic habit with square or
octagonal cross-section and a pyramid which is sometimes
truncated by the basal pinacoid (Pl. 36); also coarsely
radiating, granular, sometimes even compact.
IDENTIFICATION crystal form; fragments are not readily distinguishable from associated grossular.

Idocrase from dolomite blocks from Monte Somma, Vesuvius, described as 'Vesuvian gemstones', were recognized as a distinct mineral by Werner. Excellent crystals with distinct crystal faces from Vilui River, Yakutia, Siberia, were known as *viluite*. Occasionally used as gemstone (faceted or cabochon).

Formation and occurrence. Idocrase occurs in contact-meta-morphic limestones with grossular, epidote and wollastonite.

PLATE 46   *Feldspar: 3 – plagioclases*
Above Twinned crystals of albite, developed according to the pericline law (*cp.* Fig. 25, p. 246), with characteristic dullness. Rauris valley, Salzburg, Austria. *Below left* Labradorite. Ylöjärvi, Finland. *Below right* Labradorite with twin lamellae. Labrador, Canada.

USSR – sharp, perfect crystals (!!!) in greyish matrix, Vilui River, Yakutia, Siberia. MEXICO – Lake Jaco, Chihuahua, large rude prisms (!) to 15 cm. long, and recently, another locality nearby produced smaller but much superior crystals, some like the Siberian. CANADA – yellowish rude crystals (partly gem quality, known as 'laurelite') from near Sixteen Island Lake, Laurel, Argentueil Co., Quebec. USA – large crystals with quartz and epidote, near Sanford, York Co., Maine; crystals (!!) in asbestos deposit at Eden Mills, Lamoille Co., Vermont; the purplish variety 'cyprine' occurs massive at Franklin, New Jersey zinc deposit; massive gem quality at Pulga, Butte Co. and at various points in Siskiyou Co., California. NORWAY – crystals (!!) near Kristiansand and Oslo districts. ITALY – beautiful rich green crystals (!!) in Val Ala; dark green in ejected blocks from Monte Somma, Vesuvius.

PREHNITE

CHEMISTRY $Ca_2Al_2(OH)_2(Si_3O_{10})$, calcium aluminum hydroxyl silicate.

STREAK colorless   ORTHORHOMBIC   H $6\frac{1}{2}$ SG 2.9
COLOR usually pale green, also yellow, nearly white.
LUSTER vitreous to pearly on fracture surfaces.
PROPERTIES minute cleavage surfaces exposed on fractured material; brittle but fairly tough when compact; translucent.
CRYSTALS very small, rare, as squarish individuals; mostly radiating fibrous crusts and masses.
IDENTIFICATION typical color and crusts.

Named after Dutch Colonel van Prehn who brought specimens to Europe from Africa in 1774; sometimes used as a cabochon gem material.

Formation and occurrence. Hydrothermal late mineral encrusting cavities in basalt, diabase, and other rocks; occurs also in skarn rock cavities as a very late mineral. USA – thin to thick crusts of fine color (!!!) abundant in cavities in basalts and diabases of the Eastern states, as at Watchung Mountains, Northern New Jersey, and Farmington, Connecticut and Westfield, Massachusetts; also in fine thick crusts (!!!) with apophyllite and bornite at Centreville, Fairfax Co., Virginia; the finest specimens came from quarries in and near Paterson, Passaic Co., New Jersey, particularly New Street and Prospect Park;

also good material (!!) in copper deposits of Keeweenaw Peninsula, Michigan. SCOTLAND – fine yellow, translucent crusts (!!) in basalt Renfrewshire and Dumbartonshire (some gem quality !). GERMANY – from various basalt quarries in the Rhine-Palatinate (!). SWITZERLAND – crystals (!!!) at Muota Naira, Upper Val alps. FRANCE – excellent spheroidal masses (!!!) Bourg d'Oisans, Dauphiné.

AXINITE

CHEMISTRY $(Ca, Mn, Fe)_3Al_2(OH)(BO_3)Si_4O_{12}$, complex silicate containing hydroxyl and borate groups.

STREAK colorless   TRICLINIC   H $6\frac{1}{2}$–7   SG 3.3

COLOR mostly some shade of brown; also greyish, bluish-grey, violet-grey, and very rarely, orange; strong pleochroism.

LUSTER vitreous.

PROPERTIES uneven to conchoidal fracture; brittle; transparent to nearby opaque.

CRYSTALS sharp-edged wafer-like, in clusters with sharp edges up; sometimes massive.

IDENTIFICATION crystal shape and color; pleochroism.

Named from Greek *axine,* 'axe', for shape of crystals. Clear crystals have been used for gems.

Formation and occurrence. High-temperature hydrothermal mineral in ore veins; in marbles at contacts with igneous bodies (skarns); in granites and pegmatites lining fissures and cavities. FRANCE – superb crystals (!!!) St Christophe, Bourg d'Oisans, Isère. SWITZERLAND – Piz Vallatscha, abundant in crystals (!!) to 3 cm. POLAND – crystals in pegmatite druses, Strzegom, Silesia. USA – large clear crystals (!!!) near Coarse Gold, Madera Co., California. MEXICO – very large crystals (!–!!) to 12 cm., Mina La Olivia, and near Gavilanes, Baja California, much of the La Olivia material is gem grade. JAPAN – lustrous brown parallel groups of crystals (!!!), often of large size, from Toroku mine, Miyazaki Prefecture, Kyushu Island.

BERYL   Plate 37

CHEMISTRY $Be_3Al_2Si_6O_{18}$, seldom without traces of iron, sodium, potassium, lithium, cesium, etc.; rarely with chromium *(emerald)*.

STREAK colorless   HEXAGONAL   H 8   SG 2.7–2.9

COLOR from colorless through many green shades, blues, yellows; also pink, orange, brown, black (due to inclusions), red; commonly color-zoned.

LUSTER bright vitreous; sometimes silky due to inclusions; also aventurescent for same reason.

PROPERTIES poor cleavage on basal plane (across prisms); uneven to conchoidal fracture; brittle but fairly tough; transparent to translucent.

CRYSTALS common, as simple hexagonal prisms with flat terminations, and more rarely, with accessory faces on ends or along prisms; also etched and corroded crystals and lumps; sometimes granular.

IDENTIFICATION crystal shape, color; difficult to distinguish from quartz but has brighter luster while quartz is more oily.

The various color varieties of beryl receive names accordingly; the vivid grass-green *emeralds* are colored by traces of chromium; *aquamarine* ('sea-water color') may be pale to medium pure blue, but more commonly it is some shade of green or bluish-green and owes its color to iron; *morganite*, named by George F. Kunz after New York banker J. Pierpont Morgan, is pink or purplish-pink, pink-orange (some orange beryls turn pink when exposed to the sun or heat); *goshenite*, named after a locality in Massachusetts, is colorless but usually contains alkali elements; *golden beryl* is yellow (a rich yellow kind from South-West Africa, supposedly radioactive, was named *heliodor*); *vorobyevite* is morganite named after a famous Russian mineralogist; *cesium beryl* or *alkali beryl* are names descriptive of chemical content.

Emeralds have been known as gemstones for a very long time; they have been found with mummies in Egyptian tombs. Numerous mine shafts were discovered during the last century inland from the Red Sea coast of Egypt at Zabara, south of Koseir, which produced emeralds from about 2000 BC. Whenever one comes across reports about 'columns of emerald' it is practically always a confusion with malachite or some greenish rock, which Theophrastus described as 'false emerald'. Pliny mentions *berillus* as being closely related to *smaragdus* (emerald), which was said to be beneficial to the eyes: he reports that

Nero was supposed to have watched the gladiatorial combats through an emerald. The origin of the word emerald is not surely known. Aristotle's *De Petris* gives us information about the healing power of 'zabargad' or 'zumurrud', as an antidote for poisons and preventing children from falling; these Arabic names can easily be recognized as derivations from the Greek *smaragdos,* 'green'. Zumurrud is also the name of one of the principal female characters from the *Arabian Nights.*

Emeralds are as valuable as ever; stones of first grade rank with rubies of equal quality. The jewel-room at Vienna has a 2,205-carat carved inkwell or jar, and a Russian crystal of 12 × 25 cm. Emeralds are now made synthetically, and all kinds of imitations are being produced.

Aquamarine is far less expensive than emerald, because it occurs in large crystals of high quality. Thus in 1910 a crystal from Marambaia, Brazil, weighing 110.5 kg. ($\frac{1}{2}$ million carats) and 48.5 cm. long was sold for $25,000. Stones totalling about 200,000 carats were cut from it. Most valuable are the rarer darker blue gems.

Beryl is the most valuable ore of beryllium metal, which is used in special alloys, and in reactor technology. A copper-beryllium alloy has the properties of steel used for tool-making, but does not spark and is non-magnetic. Beryl is rare except in granite pegmatites, although huge crystals are sometimes found in these bodies. At Ponferada, Galicia, Spain, for instance, single crystals were formerly used in columns, as door-posts and such-like. About 9,000 tons of ore beryl are produced yearly.

Formation and ocurrence. By far most beryl is found in small to large crystals in granitic pegmatites, usually 'frozen' inside masses of quartz-feldspar but sometimes free-standing in cavities and then of wonderful perfection, if not of gem quality; it also occurs in some skarns, and in some carbonate-sedimentary deposits, where it has been introduced hydrothermally (Colombia); in schists intruded by pegmatites (emeralds, Urals); present in small quantities in many greisens; very rarely in volcanic rocks (Thomas Mts., Utah).

BRAZIL – this country, especially in the states of Minas Gerais, Bahia, Paraiba, and Rio Grande do Norte, is truly 'beryl land', the many pegmatites having produced gem-quality crystals of all sizes and qualities from Colonial days to the present, and more recently, large quantities of common (ore) beryl; beryl pegmatites are found mainly around Governador Valadares and Teofilo Ottoni, Minas Gerais; practically all commercial gem-quality material is produced here. MADAGASCAR – similar pegmatite fields of very large extent occur in the central parts of this island and produce exceptional morganite (!!!) and blue aquamarine (!!!), also much ore beryl. MOZAMBIQUE – the pegmatite region inland from Lourenço Marques is noted for fine morganites. SOUTH AFRICA – intensive mining produces much small emerald in the Gravelotte area. SOUTH-WEST AFRICA – aquamarines and golden beryls in superb smaller crystals (!!!) occur in pegmatites at Kleine Spitskopje and elsewhere. RHODESIA – aquamarine and much ore beryl. EGYPT – ancient, but now uneconomical, emerald deposits in schists, around Jebel Zabara, in arid mountains between Nile River and Red Sea; mostly poor-quality stones. INDIA – ore beryl in many mica mines; also emeralds in schist; aquamarines in foothills of Himalayas.

USSR – famous emerald mines of the Takovaya River region, 90 km. NE of Sverdlovsk, were discovered by a peasant who saw green crystals enmeshed in the roots of a wind-fall tree in 1831; by next year the mines were being exploited and with emerald were found alexandrite, phenazite, and other minerals; some large stones were found but most gems were small and from fair to good quality; superb aquamarine and golden beryl crystals on matrix (!!!) from pegmatites around Mursinka in the Urals, also in Ilmen Hills; beautiful specimens of blue aquamarine in vugs in porous granite at Adun-Chilon, Transbaikalia. AUSTRIA – emerald crystals in schist Habachtal, Salzburg but unimportant for gems. COLOMBIA – emerald occurrences, known and exploited by natives prior to Spanish Conquest (sixteenth century) extend over wide area and are characterized by emerald-bearing calcite veins in black carbonate sedimentary rocks; notable are Muzo (superb dark green crystals !!!) and Chivor (fine crystals !!!); virtually all the important gem emeralds of the world come from Colombian mines which are still productive; occasionally matrix specimens (!!!) are furnished. USA – much ore beryl has been produced in the pegmatite regions of the New England

and Appalachian States, but seldom good mineralogical or gemmological crystals; emerald in schist and in vugs in schist occurs at Crabtree, Mitchell Co., and around Hiddenite (superb crystals !!!), Alexander Co., North Carolina; fine blue aquamarine crystals on Mount Antero, Chaffee Co., Colorado; small raspberry-red crystals, only several mm. in size, occur with topaz at Thomas Mountains, Utah; morganite crystals (!–!!!) in pegmatites of San Diego Co., California, notably at Pala, also Mesa Grande, and less commonly at Ramona and Rincon; at the last place a small pegmatite furnishes perfect slender aquamarine crystals (!!) to 12 cm. long and 5 mm. thick.

Beryl crystals, especially the clear gemmy types, are much prized for collections, while matrix specimens are greatly in demand.

CORDIERITE

$Mg_2Al_3(AlSi_5)O_{18}$, magnesium aluminum silicate. Usually occurs as blue grains but shows strong variation in color in different directions (pleochroism): blue in one direction and yellowish-grey at right angles to this, hence the alternative name *dichroite* – 'two-colored'. Occasionally found as thick orthorhombic prisms, as at Näversberg, Sweden. Cordierite occurs in gneiss at Bodenmais, Bavaria. Clear gem-quality grains are mined in India and Madagascar.

TOURMALINE SERIES   Plates 38 and 41

All tourmalines have identical silicate structure: $(BO_3(OH)_4 Si_6O_{18})^{23-}$. The different properties of end members of this series depend on the kind of metals present, and color is often a good indication of the chemical composition. The principal end members are:

Na-Li-Al	Lithium (or alkali) tourmaline or *elbaite,* colorless, pink or green
Na-Mg-Al	Magnesium tourmaline or *dravite,* brown to brownish-black
Na-Fe-Al	Iron tourmaline or *schorl,* black.

Since other metals may enter in the composition, a large number of variations occur, but only black tourmaline (schorl) is common.

STREAK colorless   HEXAGONAL (trigonal)   H 7–7$\frac{1}{2}$
SG 3.0–3.25

COLOR colorless, white, many shades of green from blue-green to yellow-green to olive-green (nearly opaque), less commonly pink through red, yellow, orange, purplish, blue to dark blue, and black, by far the most common hue; crystals commonly color-zoned.

LUSTER vitreous.

PROPERTIES no cleavage, but crystals tend to fracture rather evenly across prisms; conchoidal to uneven fracture; brittle; opaque to transparent, but even the 'opaque' crystals are transparent in thin splinters.

CRYSTALS predominantly rounded triangles in cross-section, sometimes regular triangles or hexagons; crystals usually striated along sides of prisms, an excellent identification feature; short to very long prismatic; also fibrous, acicular, and rarely granular massive. Tourmaline is *hemimorphic*.

IDENTIFICATION crystal shapes with side striations; on doubly terminated crystals the end faces are of different character and angular position due to hemimorphism; color; associates; hardness (harder than apatite, somewhat less hard than beryl).

The chemical composition is commonly reflected in the color, and according to color, the following varieties occur: iron tourmaline, black, *schorl;* lithium or alkali tourmalines, or elbaites, are commonly weakly to strongly colored in greens, blues, pinks, etc., a deep blue kind being known as *indicolite,* colorless as *achroite,* red as *rubellite;* dravites are predominantly brown, but some green-brown types contain admixtures of iron and alkalis.

Tourmalines have been recovered from the gem gravels of Ceylon for thousands of years, the name being derived from the Singhalese name *turamali*. However, the history of tourmaline in Europe is quite recent (eighteenth century), when the Dutch brought in stones from Ceylon which they called

PLATE 47   *Feldspar: 4 – pegmatite occurrence*
Typical pegmatite cavity lining from Epprechtstein, Bavaria, Germany. Large microcline crystals coated with a crust of 'gilbertite' (finely flaky muscovite), with smoky quartz and small blue fluorite cubes.

19 ℂ 67

'aschentrekker' or 'ash-attractors', because warmed crystals and gems tended to attract ashes and dust to themselves (due to the hemimorphic character). Schorl is an old German miners' name for black tourmaline, of uncertain derivation. Aside from a minor use of cut tourmaline wafers to measure pressures by generating an electrical charge when they are squeezed violently, the only real use is as a gemstone.

Formation and occurrence. Commonly as accessory species in schists and gneiss; in some high-temperature veins and greisens; especially abundant in granitic pegmatites; dravite occurs in metamorphosed limestones.

BRAZIL – in splendid crystals (!–!!!) from the pegmatite regions of Minas Gerais (see beryl, p. 216), and similarly from Madagascar, Mozambique, etc. MADAGASCAR – exceptional large crystals with color-zoned cores to 10 cm. diameter, often sliced into cross-sections and polished to display zoning. MOZAMBIQUE – enormous crystals of rubellite near Lourenço Marques to nearly 1 m. long. SOUTH-WEST AFRICA – fine blue and blue-green crystals from pegmatites near Rossing, etc. NEPAL – recently in clear pink crystals (!!!) from foothills of Himalayas. CEYLON – abundant rolled pebbles and crystals of dravite in gem gravels, cuttable into gems. USSR – famous occurrences in Mursinka district of Urals. ITALY – small pegmatites on Elba produced exquisite matrix specimens (!!!), including pink, green, and black-topped crystals. YUGOSLAVIA – Dobrava, Drava River, Slovenia, type locality of dravite. NORWAY – Kragerö, fine schorl crystals (!!). AUSTRIA – Spittal, Carinthia, schorl crystals to 20 cm. USA – fine green and red crystals (!!) from Mount Mica, Oxford Co., Maine, and from other pegmatites in the area; splendant schorl crystals (!!!) Pierrepont, St Lawrence Co., New York; fine green crystals (!) Gillette Quarry, Haddam Neck, Connecticut; superb crystals, often double terminated and bicolored (!!!) from Himalaya Mine, Mesa Grande, San Diego Co., California and from nearby in the Pala district (large rubellites !!!) and Ramona

PLATE 48   *Zeolites*
Above Typical 'sheaves' of stilbite. *Center* Natrolite needles with a greenish cube of apophyllite, and yellowish-white chabazite. Rio Grande do Sul, Brazil. *Below* Reddish crystals of heulandite colored by hematite. Val di Fassa, Italy.

(large olive-green crystals !!). MEXICO – large 'watermelon' color-zoned crystals fromAlamos district, Baja California. AUSTRALIA – very large, to 1 kg., well-formed dravite and and schorl crystals (!!!) Yinnietharra, Western Australia, a recent find.

Section 2 Chain- and ribbon-silicates

Amphiboles (hornblende) and pyroxenes (augite) belong to this group. Minerals of both groups resemble each other in many ways and are indistinguishable from each other in small crystals or grains by the naked eye, if the cleavage cannot be seen (Fig. 23). They are also very close chemically; amphiboles, however, take up water and fluorine whilst pyroxenes do not.

Pyroxenes

Pyroxenes are further subdivided into orthorhombic and monoclinic series. Their formulae, however complex they may be, are basically $M_2Si_2O_6$. The metal positions (M) may be occupied by Ca, Na, Li, Mg, Mn, Fe^{2+}, Fe^{3+}, Al. Further complications arise by partial replacement of Si by Al. Ti is also often present, sometimes even Cr. Extensive isomorphous replacement results in many varieties.

Monoclinic pyroxenes

diopside	$CaMgSi_2O_6$
hedenbergite	$CaFeSi_2O_6$
diallage	$Ca(Fe, Mg)Si_2O_6$ with Al, Fe^{3+}
common augite	$CaMgSi_2O_6$ with much Fe^{2+}, Fe^{3+}, some Ti, Al
fassaite	$CaMgSi_2O_6$ with predominant Al besides Fe, Ti
spodumene	$LiAlSi_2O_6$
jadeite	$NaAlSi_2O_6$
aegirite	$NaFeSi_2O_6$ with some Al, Ca, Ti

Orthorhombic pyroxenes

enstatite	$Mg_2Si_2O_6$

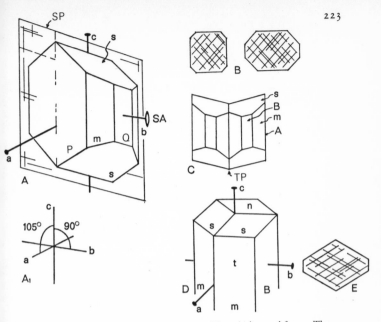

23 Pyroxenes and amphiboles. (A) Augite with typical crystal forms. The symmetry of a monoclinic crystal is here shown by the single symmetry plane SP and the single two-fold symmetry axis SA. SP divides the crystal into two identical mirror-image halves; the crystal can be rotated round SA twice (each 180°) to regain coincidence. a and b are pairs of pinacoid faces, m is a prism parallel to the *c*-axis and s is a prism oblique to the axial cross. (B) Cross-section through a rectangular and a somewhat flattened (like A) augite crystal showing characteristic octagonal outline and network of cleavage fissures at right angles to each other. (C) Augite twin, TP twinning plane. (D) Hornblende with apparent hexagonal cross-section outline; m prism parallel to *c*-axis, s an inclined prism, and n and b pinacoids. (E) Cross-section of hornblende crystal showing cleavage fissures, which, in contrast to augite, cross each other at angles of about 120°.

bronzite $(Mg, Fe)_2Si_2O_6$
hypersthene $(Fe, Mg)_2Si_2O_6$

The properties of individual members differ so much that they have to be discussed separately.

The rectangular prismatic cleavage is characteristic of all monoclinic pyroxenes, whereas the cleavage fragments of amphiboles consist of tiny prisms with diamond-shaped cross-sections. Pyroxene crystals usually have a square or octagonal

224

cross-section while the cross-section of amphibole crystals
tends to be pseudo-hexagonal (Fig. 23).

DIOPSIDE and HEDENBERGITE

Diopside shows complete miscibility with hedenbergite,
limited miscibility towards common augite. Occasionally a
little Cr or Mn.

STREAK white   H $5\frac{1}{2}$–$6\frac{1}{2}$ SG 3.3

COLOR colorless, white, grey, yellow, pale green to dark green
with increasing Fe content.

LUSTER vitreous.

PROPERTIES in crystals and coarse aggregates cleavage distinct;
fracture uneven, brittle; transparent to opaque.

CRYSTALS fully developed crystals are rare, usually obtuse-
angled, clearly monoclinic, with square cross-sections;
granular or rod-like aggregates.

IDENTIFICATION crystals and coarsely fibrous aggregates by
their square cross-section (Fig. 23) and usual bottle-green
color.

The word 'dopside' is derived from the Greek *dis,* double, and
opsis, view, because it is strongly birefringent. *Hedenbergite*
occurs either in crystals similar to dopside or coarse granular:
color, dark-green to black; streak, grey-green. The mineral was
named after the Swedish chemist who first analysed it.

Formation and occurrence. Diopside is the major constituent of
basic igneous and calcsilicate rock. Iron-rich varieties, especially hedenbergite, are typical of skarns. It is not uncommon in
lime-rich crystalline schists, and in alpine fissures in ser-
pentinite. Chrome-diopside is found in olivine nodules of
basalts.

Diallage is by its composition a definite intermediate member
towards common augite

ITALY – green, transparent crystals (!!!) sometimes of gem
quality, Val Ala, Piedmont; in limestone block ejecta, Monte
Somma, Vesuvius. SWEDEN – large crystals (!!) in skarn-
magnetite, with hedenbergite (!!!) sharp and fine crystals,
Nordmark quarry, Värmland. SWITZERLAND – crystals (!!) to
2 cm., Pizzo Cervandone; Castione, Ticino large crystals in
marble. AUSTRIA – crystals (!!!) Zillertal, Tyrol. USA – beautiful green crystals (!!!) partly gem quality, near Richville, St

Lawrence Co., New York; crystals (!!) of 'jeffersonite' (very large) and 'schefferite' at Franklin-Ogdensburg, Sussex Co., New Jersey. CANADA – fine crystals (!–!!!) at numerous places in Grenville marbles, as at Huddersfield, Quebec, Cardiff, Haliburton Co., Ontario, etc. FINLAND – chrome diopside crystals (!!!) in sulphides, partly cuttable, Outokumpu. BURMA – rolled gem pebbles of green diopside in gem gravels (!!).

COMMON AUGITE   Plate 39

STREAK greyish-green   H 5½–6   SG 3.4
COLOR dark green to black (high Fe content).
LUSTER vitreous.
PROPERTIES good cleavage; uneven fracture; brittle; opaque.
CRYSTALS common; with characteristic octagonal cross-section and skew terminations, often doubly terminated (Fig 23), frequently twinned, granular, in idiomorphic grains, also radiating.
IDENTIFICATION crystals distinct, but tiny fragments in rocks can only be distinguished from hornblende if the nearly rectangular cleavage shows up under a lens.

Augite comes from the Greek *auge*, 'luster', probably in allusion to the shiny cleavage planes as seen on fractured surfaces of augite-bearing rocks. Descriptions of these striking crystals, or of their habit at least, appeared during the eighteenth century under the name 'schorl'.

Formation and occurrence. Common augite is a frequent constituent of basalts, dolerites (diabase), andesites, phonolites, and good crystals also occur in corresponding tuffs. It has a tendency to alter to hornblende; in dolerites the augite is often completely altered to hornblende.

Fassaite and omphacite are Al-rich, therefore lighter-colored (green) varieties of common augite and are translucent. Fassaite sometimes occurs as well-developed small crystals near the contact with marble, as in the Val di Fassa, Dolomites, and at Monte Somma, Vesuvius, in limestone fragments.

The same mineral occurring as grains in highly metamorphosed crystalline schists is called *omphacite* (from the Greek *omphax,* green grape). Together with garnet, omphacite is the chief constituent of eclogite (as at Hof, Upper Franconia; Saualpe, Carinthia; Langenfeld, Ötztal, Tirol).

ITALY – sharp crystals (!!) at Bufaure, Tyrol, and to 2 cm. size (!!) doubly-terminated, on Mount Vesuvius, Naples and Mount Etna, Sicily. GERMANY – in the volcanic debris of the Eifel region, as at Firmerich, near Dank, Laacher See; beautiful crystals (!!) in limburgite at Limburg am Kaiserstuhl, Baden-Württemberg. FRANCE – crystals in the volcanic debris of the Auvergne. USA – good crystals (!) in basalt at Cedar Butte, near Tillamook, Tillamook Co., Oregon.

SPODUMENE   Plate 39

STREAK white   H 6½–7   SG 3.2

COLOR usually greyish-white, dull, but *kunzite* is lilac or purple, *hiddenite* green.

LUSTER vitreous; pearly on cleavage faces of ordinary material.

PROPERTIES perfect cleavage; fracture uneven; brittle; transparent to translucent.

CRYSTALS common, especially cleavage fragments, often very large, platy prismatic, with distinct vertical striations, also radiating; kunzite usually very etched or corroded.

IDENTIFICATION gem varieties by color; cleavage fragments of kunzite corroded and rounded.

Spodumene derives its name from the Greek *spodios,* ashen-colored. Kunzite is named after George F. Kunz, gem expert of New York City, and hiddenite after W. E. Hidden, an American mineralogist. Spodumene is an important source of lithium, which is used in chemicals and in certain alloys.

Formation and occurrence. Spodumene, lepidolite and tourmaline are the principal Li-species of complex granitic pegmatites, with spodumene being almost exclusively a pegmatite mineral although it also occurs in veins and fissure-fillings derived from pegmatites. USA – magnificent kunzite crystals (!!!) from Pala Chief mine and in the San Pedro mine (corroded fragments !!) as well as elsewhere in the Pala district, San Diego Co., California; much of the material is gem quality; large crystals of common spodumene in pegmatites of Newry, Oxford Co., Maine; beautiful gem crystals of emerald-green hiddenite (!!!) in veins in gneiss around Hiddenite, Alexander Co., North

Carolina; enormous spodumene crystals, Etta mine, Black Hills, South Dakota (record crystal of 13 m. !). BRAZIL – much kunzite and yellowish and pale green gem material from pegmatites in Minas Gerais. MADAGASCAR – fine kunzite from pegmatites.

Jadeite is a metamorphic mineral sometimes forming veins and lenses of considerable size and then quarried like rock for the sake of the ornamental material it provides to skilled Chinese carvers. The principal occurrence is near Tawmaw in Upper Burma in serpentinites, from which deposits gem jadeite has been supplied to China since the seventeenth century. Most is white to faintly greenish but it also occurs in a vivid emerald-green hue ('imperial jade') much prized for jewelry and ornaments; also in shades of red, yellow, orange, brown, and a rather pale lavender; a black variety is known as *chloromelanite*. Other occurrences are in Val di Suza, Piedmont, Italy, in San Benito Co., California, and in Japan (rather good material). The other principal jade stone, *nephrite*, a variety of actinolite, is distinguished from jadeite by having a finer texture and refusing to melt into a glassy globule in a hot blowpipe flame.

Aegirine is a typical constituent of nepheline syenites and related pegmatites and lavas. The crystals usually appear as elongated four-sided or flat, vertically striated needles, with steep pyramidal terminations; also spiky, radiating, or finely fibrous, and then very difficult to distinguish from dark actinolite. Langesundfjord, Norway, in pegmatites related to nepheline syenites; Khibin Hills on the Kola peninsula (see apatite, p. 188) and in the Ilmen Hills; Magnet Cove, Garland Co., Arkansas.

RHODONITE

CHEMISTRY (Mn, Ca)SiO_3, manganese calcium silicate; iron substitutes for manganese as does zinc (Franklin, N.J.).

STREAK colorless   TRICLINIC   H 6   SG 3.65

COLOR mostly pale to deep pink; also red, brownish-red, red-brown.

LUSTER vitreous.

PROPERTIES perfect cleavages nearly at right angles; uneven fracture; brittle; translucent to transparent.

CRYSTALS square tabular to box-like prisms but rare; mostly granular massive.

IDENTIFICATION characteristic color, cleavages; hardness as compared to similar pink rhodochrosite (latter readily attacked by hydrochloric acid but not rhodonite).

Named after the color, Greek *rhodon,* 'rose'. A calcium-rich variety is known as *bustamite,* and *fowlerite* when zinc replaces manganese. The massive material is used for ornaments and cabochon gems, and small clear crystals from Broken Hill, New South Wales are sometimes faceted.

Formation and occurrence. Formed during metamorphism of sedimentary manganese ores; in contact metasomatic deposits; a hydrothermal mineral in some ore veins. USSR – known as *orletz* in Russia, once highly prized for ornamental material and quarried near Orsk. USA – classic box-like and tabular pink crystals (!!!) to 20 cm. from marble at Franklin, Sussex Co., New Jersey; massive material from California and elsewhere. SWEDEN – small wedge-shaped bright crystals (!!) near Pajsberg, Värmland. AUSTRALIA – beautiful brownish-red square prisms (!!!), often completely transparent, in coarsely granular galena, to 5 cm. long, Broken Hill, New South Wales.

Enstatite, bronzite and hypersthene form a complete solid solution series. It is hard to distinguish between these minerals, especially as they may be intergrown with closely similar diallage. Alteration to serpentine is common.

Enstatite is sometimes present in plutonic rocks. Large crystals from Bamle, Norway; enstatite rock from Kraubath, Styria (Pl. 39).

Bronzite: the flaky-fibrous parting and bronzy schiller is very noticeable. Streak, white; hardness $5\frac{1}{2}$–6; specific gravity 3.3; color brown, green, bronze. Only occurs as elongated platy grains, not as well-formed crystals. Bronzite is a major constituent of norites. Pseudomorphs of serpentine after bronzite, retaining the flaky appearance, are called *schiller-spar.*

Hypersthene shows distinct coppery-metallic luster on parting planes; dark to black, with yellow to brown streak. Occasionally as simple orthorhombic prisms with flat pyramidal terminations, mostly as cleavage fragments or foliated-granular. Hypersthene is a constituent of certain norites and gabbros, such as those from Labrador. Found also in cordierite gneiss at Bodenmais, Bavarian Forest.

Amphiboles

Amphiboles, like the pyroxenes, can be sub-divided into a monoclinic and an orthorhombic series. They can be repsented by the basic formula $Ca_2(Mg, Fe)_5(OH, F) (Si_4O_{11})_2$.

Monoclinic amphibole series

Actinolites:
tremolite	CaMg, little or no Fe
actinolite	CaMg, increasing Fe

True amphiboles:
hornblende	CaMg, some Na, K, rich in Fe^{2+}, Al
basaltic hornblende	as above, but rich in Fe^{3+}

Alkali amphiboles:
glaucophane	Na, Mg, Al
arfvedsonite	Na, Ca, Fe^{2+}, Fe^{3+}, Mg, Al

Rhombic amphibole series
anthophyllite	$(Mg, Fe)_7Si_8O_{22}(OH)_2$

This list includes only the most important members of the two amphibole series. For the most part, instead of giving the exceedingly complicated chemical formulae, only the predominant cations are shown. An obvious difference from the augites is the presence of OH and F. There is also a distinct tendency towards bladed to acicular crystals, resulting ultimately in the formation of fine asbestos fibres.

ACTINOLITE   Plates 40 and 32

Complete miscibility from tremolite to actinolite, with considerable Fe substituting for Mg, nearly always some Na and Al.

STREAK white   H $5\frac{1}{2}$–6   SG 2.9–3.3
COLOR colorless, white, pale to dark green, depending on Fe content.
LUSTER vitreous
PROPERTIES perfect cleavage (see Fig. 23), uneven to fibrous or splintery fracture; brittle, some compact types exceedingly tough; transparent to opaque.
CRYSTALS fully developed crystals are rare, usually as prisms with no terminal faces and rhomb-shaped or hexagonal

cross-section. From coarsely radiating to fine-hairy, matted to very compact.

IDENTIFICATION white to green color, association, cleavage.

Tremolite was incorrectly named after Val Tremola in the St Gotthard massif; its true type locality is Campolungo, Ticino, Switzerland. Finely fibrous to asbestos-like actinolite (from the Greek *aktis,* a ray) is also known as *amianthus;* Dioscorides (about AD 60) described it under this name, and he also mentions its use in making fire-proof cloth. *Byssolite* is hairlike actinolite, found in alpine fissures. *Nephrite* is thoroughly compact actinolite, very widely used in Neolithic tools (in New Guinea even well into recent times). As an ornamental stone it is considered a jade and is much revered by the Chinese, who employed it in carvings for several thousand years BC. Its name derives from the Greek *nephros,* kidney, and it was worn as an amulet against kidney ailments.

Formation and occurrence. Actinolite is commonly formed during alteration of augite-bearing rocks, also in crystalline schists, and in contact marbles. It is commonly associated with serpentine, talc, chlorite, epidote, calcite, dolomite, zoisite. AUSTRIA – common as actinolite schists, often forming large clusters of sheaves in the Hohe Tauern, Zillertal Alps, at Schwarzsee with talc, and near Obergurgl, Ötztal Alps; byssolite (!!) on the Knappenwand; Gummern, Carinthia, large temolite crystals in marble. SWITZERLAND – Campolungo, Ticino, in dolomite marble; Poschiaro, Graubünden, *nephrite* for *objets d'art* obtained as a by-product from talc quarrying. ITALY – in Appenines near La Spezia, *nephrite.* USSR – Khara-Zelga brook, Lake Baikal, *nephrite* (!!!). CHINA – Kuen-Lun Mountains, *nephrite* (!!!). USA – Jade Mountain, Alaska, *nephrite* (!!); nephrite (!!!) in southern regions of Wyoming; at several places in California. NEW ZEALAND – excellent nephrite (!!!) originally used by the Maoris.

Hornblende group Plate 40

STREAK brown   H 5½   SG 3.2 (lower than augite)
COLOR greenish-black to black.
LUSTER vitreous.

PROPERTIES cleavage distinct; fracture uneven; brittle; opaque.
CRYSTALS well-developed crystals not common; usually with few forms only (Pl. 40, Fig. 23); not infrequently twinned, spiky, fibrous, granular.
IDENTIFICATION well-developed crystals easily recognizable by form, otherwise difficult to distinguish from actinolite and augite.

Formation and occurrence. Members of the group are rock-formers and very widespread in syenite, diorite, hornblende granite, trachyte, phonolite, basalt, and corresponding tuffs. Common hornblende is the major constituent of amphibolites. Some varieties also occur as contact-metamorphic minerals. AUSTRIA – Kuruzzenkogel near Fehring, Styria, basalt, horn-blende; Leiser gorge near Spittal, Carinthia, crystals with clinozoisite in eclogite. NORWAY – Arendal. CZECHOSLOVAKIA – Bilina, Bohemia, crystals (Pl. 40). GERMANY – Laacher See, Eifel. USSR – Falkenberg, N. Urals, crystals up to $\frac{1}{2}$ m. long in gabbro pegmatite.

GLAUCOPHANE

Translucent, blue-grey to blue-black. Streak blue-grey, nearly always fibrous to granular. In crystalline schists, as at Val de Bagues, Valais, Switzerland. *Arfvedsonite,* deep blue-black with dark blue-grey streak, e.g. Langesundfjord, Norway. *Croci-dolite,* a Na-Fe^{3+} amphibole, occurring N. of Prieska, along the Orange River, S. Africa, as an easily spinnable asbestos; silicified pieces are tiger eye (see quartz, p. 133). *Anthophyllite,* almost invariably lath-like to fibrous, is clove-brown and may resemble bronzite and hypersthene. In crystalline schists, e.g. at Bodenmais, Bavaria, or at Falun, Sweden. Glaucophane is the blue-grey or greyish-green mineral which with natrolite forms the matrix rock at the famous benitoite deposit in San Benito Co., California.

WOLLASTONITE

$CaSiO_3$, triclinic, usually coarsely to finely fibrous, also radiating, with vitreous to silky luster. A typical contact-metamorphic mineral in limestone.

Section 3 Sheet silicates

Nearly all the minerals belonging here are micas or are mica-like in character.

TALC (STEATITE) and PYROPHYLLITE

CHEMISTRY talc $Mg_3Si_4O_{10}(OH)_2$, magnesium hydroxyl sili-
cate, with a little Fe and Al; pyrophyllite $Al_2Si_4O_{10}(OH)_2$.
STREAK white   MONOCLINIC   H 1   SG 2.8
COLOR colorless, white, yellowish, greenish, brown.
LUSTER greasy to strongly pearly, soapstone with a dull sheen.
PROPERTIES highly perfect cleavage; hackly fracture; soft, feels
greasy; translucent to opaque.
CRYSTALS extremely rare, apparently hexagonal flakes, coarsely
to finely foliated; *soapstone* is a compact variety of talc.
IDENTIFICATION hardness, ease of rubbing off flakes, color often
close to sericite (see muscovite, below). Talc in hot cobalt
nitrate solution becomes red, pyrophyllite and sericite
become blue.

Pyrophyllite is very similar to talc. Talc is an ancient Arabic
word, but the 'talc' in Aristotle's *De Petris* is mica, to judge by
its description. *Steatite* (from the Greek *stear*, tallow) is mentioned by Pliny and Theophrastos. Talc, soapstone and pyrophyllite are all used in large quantities. Soapstone can be cut or
turned with ease and becomes extraordinarily hard on heating,
so that many heat-resistant utensils, burners and vessels,
insulators and fire-proof bricks are made with it. Powdered talc
is used as a filling for paper and rubber, in paint-making, and
for several other purposes. In China talc and pyrophyllite were
favorite carving materials. Very pure talc provides the basis of
cosmetic powder.

Formation and occurrence. Talc is formed during the hydrothermal alteration of other rocks. Pseudomorphs after other
species are common. Low-grade metamorphic talc schists
contain dolomite, actinolite, magnetite. GERMANY – Göpfersgrün, Fichtelgebirge, Bavaria, where granite, amphibolite,
dolomite and other rocks are partly or completely altered to
soapstone, numerous pseudomorphs (!!!) after quartz crystals
and other minerals. NORWAY – talc pseudomorphs (!!!) after

enstatite occur at Bamle and Snarum. AUSTRIA – beautiful coarsely foliate specimens (!!) Greiner Alp, Zillertal. USA – fine foliated talc in large plates Holy Springs, Cherokee Co., Georgia: fine pyrophyllite in radiating star-like masses in several deposits in North Carolina, at Graves Mountain, Lincoln Co., Georgia, and Indian Gulch, Mariposa Co., California.

Micas

The basic formula of the micas is $MAlSi_3O_{10}(OH)_2$. Individual members are formed when various combinations of metals enter the M position, and F may partly replace OH.

muscovite (K-mica)	$KAl_3Si_3O_{10}(OH)_2$
paragonite (Na-mica)	Na, K, Al, no F
phlogopite (K-Mg-mica)	K, Mg, abundant F
biotite (K-Mg-Fe-mica)	$K(Mg, Fe)_3AlSi_3O_{10}(OH)_2$
zinnwaldite (Li-Fe-mica)	K, Li, Al, Fe
lepidolite (Li-mica)	K, abundant Li, Al, F
glauconite	K, Ca, Na, Fe, Al, Mg, no F
margarite (Ca-mica)	Ca, Al, no F

Complete miscibility does not exist. Micas are monoclinic, while crystals have hexagonal or orthorhombic cross-sections. All micas have highly perfect cleavages, and thinnest flakes are flexible.

MUSCOVITE   Plate 41

CHEMISTRY potassium mica, often with Na, Fe.

STREAK white   MONOCLINIC   H 2–2½   SG 2.8

COLOR colorless, yellowish, greenish, reddish. *Fuchsite* is bright green, due to chromium.

LUSTER pearly, but also silvery metallic, especially in tiny weathered flakes.

PROPERTIES perfect basal cleavage, very easily developed; flakes flexible and elastic; difficult to fracture; transparent but compact varieties opaque.

CRYSTALS uniformly well-developed tabular or prismatic crystals are rare; usually tabular with irregular outline, often quite large; from scaly to compact as so-called 'sericite'.

IDENTIFICATION excellent cleavage; color; crystal shape

(pseudohexagonal); phlogopite (very similar) is soluble in hot sulphuric acid, but muscovite is not.

Mica is derived from the Latin word for 'crumb'. Pale or colorless muscovites were named after Moscow, because large quantities of 'muscovy' or 'Russian glass' used to be sent to western Europe for use as window panes.

Muscovite and the closely related phlogopite are of considerable technological importance to the electrical industry for dielectrical applications, as fireproof roofing material, in certain paints, etc.

Formation and occurrence. Muscovite is exceptionally well developed in granite pegmatites. *Sericite,* the fine-grained variety, found in phyllites, mica schists, and also in marbles. Muscovite does not occur in extrusive rocks, but can nearly always be found as flakes in sandstones and clays. Good muscovite crystals for collectors are surprisingly rare and obtained only from open cavities in pegmatites, although some large plates from pegmatites are collected for their inclusions of flattened almandite garnet or manganese-iron dendrites. USA – fine green crystals (!!) in Rutherford mines, Amelia, Amelia Co., Virginia on snow-white albite, also delicate pink crystals of small size; curved compact, or 'ball' mica, from Branchville, Fairfield Co., Connecticut; fine garnet and dendrite inclusions (!!) in various mica mines of the Piedmont district, North Carolina. BRAZIL – fine euhedral crystals (!!) from pegmatite cavities, sometimes twinned into six-pointed 'stars'. USSR – crystals (!!) with topaz Alabashka, Mursinka, Urals. INDIA – an enormous crystal of 3 × 5 m. and weighing 85 tons was found at the Inikurti mine, Nellore, a world's record.

PARAGONITE

Monoclinic sodium mica, usually occurring as flaky masses of very small silvery crystals; the notable crystals of blue kyanite and brown staurolite from Pizzo Forno, Tessin, Switzerland are imbedded in paragonite schist.

PHLOGOPITE

CHEMISTRY potassium-magnesium mica.
STREAK colorless   MONOCLINIC   H $2\frac{1}{2}$-3   SG 3

COLOR red-brown to brown, yellow-brown.

LUSTER commonly with strong coppery semi-metallic luster on cleavage plane.

PROPERTIES like muscovite; translucent to transparent.

CRYSTALS pseudo-hexagonal outline, common, small to very large; also in aggregates of coarse flakes or crystals.

IDENTIFICATION crystal form; color; associated minerals.

Named after the Greek, *phlogopos*, 'fire-like', in regard to red-brown color; an important mica sometimes found in large deposits.

Formation and occurrence. In marbles; in magnesium-rich pegmatites; in skarns. CANADA – sharp crystals (!!) in Grenville rocks of Ontario and Quebec; world's record crystal from Lacy mine, Ontario, 90 tons, measuring 10 × 4.5 m. USA – sharp prismatic crystals (!!!) in marble of the zinc deposits of Franklin and Ogdensburg, New Jersey. MADASGASCAR – crystals (!!!) in marble with diopside, Fort Dauphin region.

BIOTITE

CHEMISTRY potassium-magnesium-iron mica.

STREAK white   MONOCLINIC   H $2\frac{1}{2}$–3   SG 3.3

COLOR always colored; usually very dark brown to black, even the thinnest flakes are distinctly brown, or red in *rubellane* from lavas and tuffs; rarely green.

LUSTER pearly to golden-metallic when weathered.

CRYSTALS well-defined crystals rare; mostly as irregular tablets, flaky, compact form unknown.

IDENTIFICATION cleavage, dark color; associates.

Biotite was named in honor of the French physicist and chemist Jean Baptiste Biot (1774–1862). Gold-colored weathered biotite is known as 'cat-gold' in Germany.

Formation and occurrence. Biotite, the commonest of the micas, is found as a rock-constituent in granites, diorites, trachytes, phonolites, etc.; it is also very widespread in crystalline schists, although less common than muscovite. Large crystals occur in pegmatites and certain contact-metamorphic rocks. Biotite alters easily; in the presence of sea water *glauconite* is formed (tiny green grains and flakes in marine sedimentary rocks). The large flakes and strip-like flakes found along borders of granitic pegmatites are biotite. ITALY – lustrous crystal groups (!!) at Monte Somma, Vesuvius.

ZINNWALDITE

A lithium-iron mica of brown color, named after the occurrence in the Zinnwald, Erzgebirge, Germany; occurs in greisens and some granitic pegmatites.

LEPIDOLITE   Plate 41

Lithium mica; common only in granitic pegmatites and some greisen deposits of tin-tungsten. It is often grown upon muscovite crystals (Pl. 41) as a result of substitution of lithium for potassium, and then forms beautiful lace-like rims. Typically pale pink, purplish-pink, or sometimes pale purple, and associated with gem tourmaline, beryl, topaz, etc. Distinguished by coloring hot blowpipe flame with crimson. USA – superb crystallizations (!!!) at Himalaya mine, Mesa Grande and in enormous blocks enclosing pink tourmaline at Stewart Lithia mine, Pala, San Diego Co., California; abundant in New England pegmatites. GERMANY – large cleavages from Penig, Saxony. MOZAMBIQUE – large curved masses of 'ball' lepidolite from Zambesi region. From numerous localities in Brazil and Madagascar.

MARGARITE

Rare. Cleavage plates are not elastic (hence its name of 'brittle mica'). Silvery color; forms 'muffs' around corundum crystals in gneiss and schist type deposits; beautiful veinlets in emery ore at Chester, Massachusetts; also with corundum in North Carolina deposits.

CHLORITES   Plates 13, 16, 45

Complete mixed crystal formation exists between all members of the group. It is therefore almost impossible to effect a more precise identification beyond 'chlorite' without further detailed examination. A few important members were given fashionable names in addition to an endless number of varieties, which are of local interest only. The basic chlorite formula is $(Mg, Fe, Al)_6 (Al, Si)_4 O_{10} (OH)_8$. The principal groups result from Fe and Al entering into the composition; further varieties arise out of Cr, Ni, Li, Mn etc. content. *Penninite,* Si-rich and Fe-poor, occurs as tiny hexagonal prisms in alpine fissures. *Clinochlore* and *prochlorite* are more Fe- and Al-rich, but contain less Si. *Prochlorite* is the chief constituent of chloritic schists. *Chamosite* and *thuringite* are Fe-rich chlorites found in limonitic ores (Minette, Lorraine), and sometimes form economic iron-ore deposits.

CHLORITE

STREAK white to green   MONOCLINIC   H $1–2\frac{1}{2}$ SG 2.5–2.9, depending on Fe content

COLOR green, ranging from nearly colorless to black.

LUSTER vitreous, pearly to dull.

PROPERTIES very good cleavage, but less perfect than in micas. Individual flakes not elastic, brittle.

CRYSTALS rare, pseudohexagonal tablets and prisms, coarsely flaky to compact, frequently as a coating on other minerals.

IDENTIFICATION green color, cleavage; practically impossible to recognize variants without advanced techniques; serpentine is harder and tougher.

Chlorite was named after its color (from the Greek *chloros,* green), penninite after the Pennine Hills in England, thuringite after Thuringia in Germany, chamosite after Mt Chamoson, Switzerland. Chlorite is sometimes used as a reflective green pigment in wallpaper.

Formation and occurrence. Common chlorites are formed hydrothermally or as low-grade metamorphic alteration products of basic plutonic rocks. Pyroxenes and amphiboles may also alter to chlorites. Chlorite is a major constituent of alpine fissures, though nearly always as a powder or dusting on quartz and feldspar crystals etc. AUSTRIA – in fissures in the Hohe Tauern; Pusygraben near Lölling, Carinthia, encrustations of chlorite with magnetite, surrounding large crystals of almandine. SWITZERLAND – very widespread in fissures and cavities; particularly upon rock crystal, adularia, sphene; Rümpfischwäng near Zermatt, large penninite crystals (!!) with idocrase, demantoid garnet, amphiboles, diopside, magnetite etc. ITALY – Mussa Alp, Ala valley, Piedmont, curved crystals (!) with grossular and diopside. CANADA – Bagot, Renfrew Co., Ontario, large cleavage flakes of clinochlore. USA – superb large crystals (!!!) in Tilly Foster iron mine, Brewster, Putnam Co., New York.

SERPENTINE (ANTIGORITE and CHRYSOTILE)
Plate 42

Serpentine occurs in two structural modifications: as lamellar antigorite, in accordance with the formula of a sheet silicate; or as fibrous chrysotile, in which the sheets have curled up into tubes.

CHEMISTRY $Mg_6Si_4O_{10}(OH)_8$, magnesium hydroxyl silicates, always with some Fe, sometimes with Ni, Cr.

STREAK white   MONOCLINIC   H $2\frac{1}{2}$–5   SG 2.6

COLOR white to green, yellow, yellowish-brown, reddish-brown to black; chrysotile yellow, shimmering.

LUSTER dull to waxy; chrysotile with strong silky luster.

PROPERTIES *antigorite* platy cleavage, like chlorites, but seldom distinct; *chrysotile asbestos* separates into very thin fibers, tough.

CRYSTALS unknown; *antigorite* always compact, *chrysotile* can be compact or fibrous.

IDENTIFICATION color; compact very tough and harder than chlorite; chrysotile typically fibrous with greenish or golden color and is soluble in acids, in contrast to *amphibole asbestos* which is less finely fibrous and tends to be darker in hue (greens, blues).

Serpentine was named because of flecked green markings, resembling snakeskin; it was believed to be effective against poisons and snake-bites. Antigorite was named after its occurrence in the Val di Antigorio, Piedmont, Italy; and chrysotile from the Greek *chrysos*, gold, and *tilos*, fiber.

Because of its toughness, serpentine is suitable for carvings. Creamy-yellow, translucent gem serpentine is particularly useful for this kind of work.

'Asbestos' is a general term for all finely fibrous minerals; 95 per cent of asbestos is in fact chrysotile, although crocidolite, anthophyllite and hornblende also occur as asbestos. Tough and elastic 'mountain leather' and 'mountain cork' are finely fibrous asbestos, mainly chrysotile. Long spinnable fibers are used for making fire-proof yarn and cloth, while other varieties, mixed with cement, are made into roofing and floor tiles, and most of all, into brake-linings.

Formation and occurrence. Serpentine is formed during hydro-thermal-metamorphic decomposition of basic igneous rocks, and of limestones and dolomites. Occasionally pseudomorphs of serpentine, after olivine, bronzite, etc., are found. Serpentine rock is widespread. AUSTRIA – Bernstein, Burgenland, precious serpentine for *objets d'art;* Griesserhof near Hirt, Carinthia, serpentine for ornamental purposes, containing chrysotile with magnetite and dolomite. SWITZERLAND – Kemmleten, southern Hospental, Uri, serpentine for decorative work. ENGLAND – the Lizard, Cornwall, fine colored decorative serpentines.

USA – yellow and green gemmy serpentine near Montville, Morris Co., New Jersey; pseudos after amphiboles in marbles of St Lawrence and Orange counties, New York; gem quality (!!!) *williamsite,* colored vivid green by chromium, in chromite mines in region around Rock Springs, Pennsylvania; 'mountain leather' in serpentine body near the Dallas Gem mine, San Benito Co., California; chrysotile asbestos (!!!) fine golden, in Salt River Valley, Arizona. CANADA – chrysotile asbestos (!!) in seams in impure serpentine, around Asbestos, Black Lake, Thetford Mines, Quebec.

Clay minerals

Clay minerals are practically always decomposition products of other minerals. Micas can take up water and alter to hydrous micas, such as *hydromuscovite,* which occurs in compacted clays. *Illite* is the chief constituent of marine clays. *Montmorillonite,* a hydrous Al-Mg-Na-silicate, is an important soil clay, as it has the capacity to absorb water; it also forms its own rocks, such as bentonite or bole. During swelling the lattice planes separate from each other and water (and also oils, pigments, gases, or other substances) intrudes between the layers. Because of this property bentonite has been used for centuries as an intestinal remedy. The *kaolinite* group includes three species with the common formula $Al_4Si_4O_{10}(OH)_8$. *Kaolinite* is the chief constituent of china clay and is formed largely by weathering or thermal decomposition of feldspars. *Dickite* and *nacrite,* the other two minerals of the group, are also formed hydro-thermally and, uncharacteristically, occur in recognizable crystals. *Sepiolite* (meerschaum), a hydrous magnesium silicate quarried at Eskisehir in Anatolia, is used for making tobacco pipes.

Interesting white kaolinite pseudomorphs after twinned feldspar crystals occur in the china clay pits of St Austell, Cornwall, England; pink montmorillonite forms pseudos after pink tourmaline in the gem pegmatites of Pala, San Diego Co., California.

Section 4 Framework silicates

NEPHELITE

CHEMISTRY $KNa_3(AlSiO_4)_4$, potassium-sodium aluminum silicate.

STREAK white   HEXAGONAL   H $5\frac{1}{2}$–6   SG 2.6

COLOR transparent to cloudy; colorless, white, pale grey; also yellowish, brownish, greenish, reddish.

LUSTER vitreous, massive varieties with marked oily luster.

PROPERTIES cleavage indistinct, conchoidal to uneven fracture; brittle; translucent to transparent.

CRYSTALS rare, hexagonal prisms with basal plane; mostly massive granular.

IDENTIFICATION oily luster, softer than quartz, absence of cleavage; crystals.

Nephelite comes from the Greek *nephele*, cloud, because clear crystals become cloudy in strong acids; *elaeolite* comes from *elaion,* oil, in allusion to the oily appearance of the grains.

Formation and occurrence. Nephelite is an important rock-former, mainly in nephelite-syenites and related pegmatites, also in phonolites, nephelite-basalts etc. ITALY – small glassy prisms in cavities in ejected limestone blocks, Monte Somma, Vesuvius; crystals (!) Capo di Bove, Rome. CANADA – large rude hexagonal prisms (!!) to 10 cm. diameter with sodalite at Bancroft, Hastings Co., Ontario.

ANALCIME   Plate 43

CHEMISTRY $NaAlSi_2O_6.H_2O$, sodium aluminum silicate with water, often some K, rarely Mg and Ca.

STREAK white   CUBIC   H 5–5½   SG 2.2

COLOR colorless, white grey, pink to flesh-red (hematite inclusions).

LUSTER vitreous to dull.

PROPERTIES cleavage absent, fracture uneven, brittle; transparent (rarely) to translucent.

CRYSTALS usually in well-formed individuals up to fist-size (trapezohedrons); also coarsely granular to compact.

IDENTIFICATION crystal form; fuses fairly easily, in contrast to leucite.

The name is derived from the Greek *analkis,* powerless, in allusion to weak electrical properties.

Formation and occurrence. A late hydrothermal mineral in gas cavities of basic volcanic rocks, but also as a hydrothermal product in nepheline-syenite pegmatite and other types of deposit. USA – crystals (!!), sometimes enclosing copper, in copper deposits of Keeweenaw Peninsula, Michigan; beautiful crystals (some red) in basalts of Northern New Jersey as at Prospect Park, Paterson; crystals (!!) Table Mountain, Jefferson Co., Colorado. CANADA – in the cavities in basalt at Five Islands, Cape Blomidon, and other localities, Nova Scotia. ITALY – huge crystals (to 10 cm. diam.) Seiser Alp, Tyrol; small fine crystals Cyclopean Islands, near Sicily. GERMANY – in the Harz. ENGLAND – crystals (!) Isle of Skye, Scotland. ICELAND – crystals (!!) Berufjord and Fasrudsfjord. AUSTRALIA – sharp crystals (!) Flinders Island, Tasmania and near Sydney, New South Wales.

LEUCITE   Plate 43

CHEMISTRY $KAlSi_2O_6$, potassium aluminum silicate.

STREAK white   Below 605°C TETRAGONAL, above that CUBIC   H 5½–6   SG 2.5

COLOR rarely colorless, mostly white or grey.

LUSTER vitreous to dull, earthy.

PROPERTIES cleavage absent, fracture conchoidal, brittle; translucent, rarely transparent.

CRYSTALS usually tetragonal paramorphs after cubic form; crystals invariably trapezohedrons (Fig. 4D), never massive.

IDENTIFICATION crystal form; unlike analcime, which it

resembles closely, it occurs imbedded to tuff rocks instead of in cavities.

Leukos is the Greek for 'white'. Because of its crystal form and occurrence in lavas the mineral was thought to be garnet bleached by fire.

Formation and occurrence. Almost exclusively in basic volcanic rocks, such as leucite-phonolite or leucite-basalt, if the melts are sufficiently poor in Si; leucite, therefore, never occurs with quartz, and is frequently accompanied by aegirite and nephelite. GERMANY – Laacher See, Eifel; Eichelberg near Rottweil, Kaiserstuhl, Baden-Württemberg. ITALY – in Vesuvian lavas; Capo di Bove and Alban Hills near Rome, fine crystals (!!).

Feldspars

TABLE 11   A survey of the Feldspar Groups

	monoclinic	*triclinic*
$(K, Na)AlSi_3O_8$	orthoclase sanidine adularia soda-orthoclase	microcline anorthoclase
		plagioclases
$NaAlSi_3O_8$	100% Na	albite
complete	80% Na	oligoclase
solid	60% Na	andesine
solution	40% Na	labradorite
series	20% Na	bytownite
$CaAl_2Si_2O_8$	0% Na	anorthite

Feldspars are the most abundant of all minerals, and account for nearly half the volume of the Earth's crust. The many varieties and members in this group can be referred to two basic formulae:

$(K, Na)AlSi_3O_8$ potassium feldspars with little to considerable sodium content

$(Na, Ca)Al(Al, Si)Si_2O_8$ plagioclases, or sodium-calcium feldspars

Potassium feldspars always carry some sodium in varying amounts, also a little Fe, often as admixed hematite, and occasionally Ba. In plagioclases a little K is always present, with traces of Fe, Ba, Sr.

As shown in Table 11, potassium feldspars are monoclinic or triclinic. External differences between the two are small. The two prominent cleavage directions are perpendicular to one another in monoclinic crystals, and meet at angles very close to 90° in triclinic crystals (Fig. 24). Complete miscibility between K and Na feldspars exists at high temperatures only; the separation expected during cooling, however, does not always take place or is only partial. Ordinarily orthoclase in slowly cooled plutonic rocks, and microcline in pegmatites, become wholly unmixed, with the sodic feldspar component regularly sandwiched into the potassium feldspar in bands or layers, resulting in the so-called *perthite* structure. Microcline in pegmatites clearly displays perthitic layers. Sanidine, on the other hand, is not all unmixed, and remains transparent despite a high Na content. Adularia is also fairly clear, but contains only very little Na.

Most potassium feldspars are morphologically very alike and grade into one another, making it very difficult to identify them correctly. The same can also be said about the plagioclase members, both in respect to each other and in their distinctions from potassium feldspars. The plagioclases are an excellent example of a complete solid solution series; Na^+ and Ca^{2+} replace each other in all proportions, but to balance valencies an increase in Ca must be accompanied by an increase in Al plus a decrease in Si content.

Feldspars are readily decomposed to clays in the course of normal weathering; in tropical regions weathering eventually forms bauxite. With hydrothermal activity potassium feldspar is altered to sericite and kaolinite, also to analcime and other minerals; plagioclases alter to zeolites, Ca-rich ones particularly to *saussurite,* a mixture of zoisite and scapolite, and also to epidote and other minerals.

STREAKS colorless   H 6, plagioclases to $6\frac{1}{2}$ SG potassium feldspar 2.5–2.6, plagioclases 2.6–2.8, with increasing Ca

COLOR generally pale colors; white, colorless, yellowish, greenish, reddish, reddish brown, tan, sometimes nearly black.

LUSTER vitreous to earthy in partly altered types; also silky in adularescent material (moonstone); pearly on cleavage planes; also with blue and silver sheens (moonstone), vivid

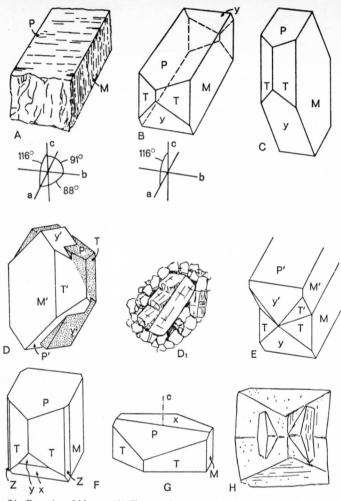

24 Potassium feldspars. (A) Cleavage fragment of a microcline crystal with perthitic structure; c and b are cleavage surfaces, with c displaying the better cleavage. The fissures upon the cleavage planes show the direction of the interlaminated sodium-feldspar (albite) in the so-called perthitic structure, which causes the irregular fracture upon the forward end. The axial cross shows the triclinic angles of microcline. (B) Orthoclase crystal, elongated along *a*-axis, with pinacoid faces c, b and y, and prism m. The monoclinic axial cross below shows the inclination of the *a*-axis, with a–b and b–c at right angles to each other. (C) Tabular orthoclase crystal. (D) Carlsbad twin like the specimen in Pl. 44, with face letters marked to distinguish the twin

color reflections (labradorescence), spangled reflections (aventurescence).

PROPERTIES two perfect cleavages intersecting at right angles or nearly so (see Fig. 24), one direction more perfect; fracture uneven and usually stepped, rarely conchoidal; brittle; translucent to transparent.

CRYSTALS common *orthoclase* normally occurs in grains or in poor to well-defined crystals in plutonic and igneous rocks. The crystals show three preferred forms: tabular on b, elongated prisms parallel to the a-axis (see Fig. 24, Pl. 44), or blunt prisms parallel to the c-axis. The main features are the pinacoids b and c, the prism m and the (usually smaller) pinacoids y. Corners are often truncated. Twinning is frequent, mainly as Carlsbad twins (Fig. 24, Pl. 44). The least well-defined rectangular crystals show a cleavage surface that is half smooth and shiny, and half irregularly broken (Fig. 24, D_1). The translucent, tabular crystals of *sanidine* often occur as Carlsbad twins, usually cracked and with rough surfaces. *Adularia* is a variety of orthoclase that occurs in alpine fissures. Single crystals like calcite rhombs; fourlings and twins are frequent, especially Manebach twins (see Fig. 24H); rectangular prisms are characteristic. Transparent to translucent crystals of adularia are commonly coated with green chlorite.

In habit and chemistry, *microcline* (Pl. 44) largely resembles orthoclase. Occasionally the triclinic symmetry is very obvious, e.g. in amazonite from Pike's Peak region (Pl. 44). Usually potassium feldspars, originally monoclinic, alter to microcline by twofold lamellar twinning. The two lamellar directions intersect at an angle very near to 90°. This lamellar lattice can only be recognized under a microscope. Potassium feldspars in

individuals. The twinning plane is irregular; the crystals are commonly grown simply parallel to b. Such twins occur in quantity in coarse-grained granites; fractured crystals then display on part of the fracture surface the smooth cleavage plane of c and the irregular fracture as shown in (D) 1. (E) Manebach twin on c. (F) Stout prismatic crystal elongated along the c-axis with an additional prism z. (G) Adularia crystal with seemingly orthorhombic faces (Pl. 45, upper left). (H) Adularia as a Manebach 'fourling' depicted in natural, distorted form (sketched from Parker, *Die Mineralfunde der Schweizer Alpen*, 1954).

25 Albite. (A) Albite crystal, similar to Fig. 24 C, but now with triclinic symmetry. Prisms are not possible in the triclinic system, hence the four faces of the monoclinic prism m are divided into two pairs designated M and m (pinacoids). (B) Albite crystal twinned according to the pericline law, sketched from an actual specimen (Pl. 46, upper). The vertical striations on M and m are due to the lamellar structure caused by repeated twinning according to the albite law (see Fig. 8), while the short crossways striations are cleavage fissures after c.

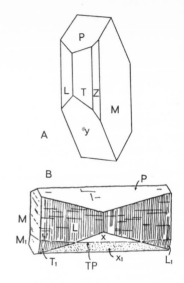

granite pegmatites are commonly described as orthoclase but in fact they are usually microcline (particularly the developed crystals in druses).

Of the *plagioclases* (Figs. 24, 25), *albite* and *anorthite* occur in crystals, the other members of the series are either massive or granular. Twin striations are nearly always distinct, especially on cleavage surfaces.

Albite (Pl. 46) occurs in tabular crystals like orthoclase (Figs. 24, 25), or elongated parallel to the *b*-axis and tabular parallel to P as in the variety *pericline* (Pl. 46, Fig. 25), in alpine fissures in association with chlorite. Twinning is common, either according to the albite law (Fig. 8), or according to the pericline law (Fig. 25). Thin blades of albite in pegmatite are known as *cleavelandite*.

The remaining plagioclases occur almost exclusively as aggregates. *Labradorite* (Pl. 46) sometimes forms large cleavage surfaces upon irregular crystals of considerable size. *Anorthite* is found now and again in small, form-rich, prismatic crystals.

Gem varieties. Moonstone is any feldspar species displaying a weak to strong silvery sheen, or sometimes a blue sheen, with a somewhat silky luster; *adularias* may also show the moonstone effect, hence the gemmologist's name, *adularescence*. *Labrador-escence* is the name given to vivid colors observed in certain directions on specimens of labradorite, notably the material from Labrador, hence the name. *Aventurescence* is a spangly effect due to small platy crystals of hematite or goethite aligned in planes within the host feldspar; the beautiful orange-spangly oligoclase (a plagioclase) from Norway is called *sunstone*. *Amazonite* is a pale to medium green or blue-green microcline found in some pegmatites. In some albites, reflected colors, like those in labradorite, but not as intense, form the variety known as *peristerite (peristerism)*.

Formation and occurrence. GERMANY – Hagendorf, Bavaria, large pegmatites (microcline crystals up to 30 m. long) with cleavelandite, phosphates and very large quartz crystals; Epprechtstein, Bavaria, well-known pegmatite druses with microcline crystals (!!) and blades of albite, in addition to smoky quartz, zinnwaldite, tourmaline, fluorite, apatite, topaz and uranium micas; Ochsenkopf in the Fichtelgebirge, good crystals of orthoclase in crystalline granite; Laacher See, Eifel, sanidine crystals (!); Drachenfels in the Siebengebirge, sanidine in trachyte (partly anorthoclase). POLAND – Strzegom (Striegau), WSW. of Wroclaw, microcline crystals (!!), smoky quartz, chabazite, stilbite, heulandite, epidote, axinite. CZECH-OSLOVAKIA – Karlovy Vary, Carlsbad twins. AUSTRIA – Hohe Tauern, adularia (!!) and pericline (!!!) in alpine fissures, e.g. at Habachtal, Griewiesalp in the Rauristal, and Knappenwand; Greiner, Rothenkopf and other places in the Zillertal Alps, pericline (!!!).

SWITZERLAND – St Gotthard massif, at numerous places between Val Cristallina and St Gotthard, adularia (!!!) first found here, at Adular, with pericline, quartz, apatite etc., adularia crystals up to 25 cm. long in the lower Val Cristallina; Piz Beverin, Graubünden, transparent multiple twins of albite. ITALY – Baveno on L. Maggiore, microcline twins (!!!) (the Baveno law); Valle di Vizze near Vipiteno in the Alto Adige, pericline; Monte Somma, Vesuvius, anorthite crystals in ejected blocks; Cyclops Is. off Catania, Sicily, anorthite. NORWAY – Larvik, augite-syenite with labradorescent sodic orthoclase; Tvedestrand, sunstone (!!). FINLAND – Ilijärvi near Ylämaa in SW. Finland, specially fine labradorite showing full play of colors; this is the 'spectrolite' of commerce. USSR – Miass in the Ilmen Mts., Urals, amazonite; Zhitomir, Ukraine, labradorite. CANADA – eastern part of Labrador, labradorite in grains up to $\frac{1}{2}$ m. in size, labradorescent (!!). USA – Sulzer, Prince of Wales Island, Alaska, adularia; Pike's Peak and the country to the west, especially Crystal Peak, Colorado, famous for amazonite (!!!) in druses, with very dark, nearly black, smoky quartz; Rutherford mine, Amelia, Virginia, gem-quality amazonite (!!!) and fine cleavelandite (!!!); San Diego Co. gem mines, microcline (!). JAPAN – Miyaki Island, Tokyo Bay, large crystals of anorthite (!!!).

Identification. Membership in the feldspar group is easy to establish from hardness and cleavage, but identification of individual members is more difficult. Adularia and pericline can be recognized with ease and certainly from their habit, plagioclase from its lamellar twinning if nothing else. Potassium feldspars and sodic plagioclase are not attacked by acids whereas calcium plagioclases after labradorite decompose. Orthoclase cannot be distinguished from microcline without advanced techniques, although paragenesis gives a clue. Sanidine cannot be told apart from anorthoclase in hand specimens.

Sodalite series

The cubic minerals sodalite, nosean and hauyne form a solid solution series of $Na-Ca-CO_3-SO_4$ silicates of varying composition. Sodalite is associated with nephelite in elaeolite-syenites; nosean and hauyne occur occasionally in ejected blocks at Laacher See, Germany, hauyne in blue grains in basaltic lava at Niedermendig, S. of the Laacher See.

Of this series sodalite is the most common and important species. The name alludes to its sodium content. Crystals are very rare; most sodalite is granular massive, sometimes forming large, nearly pure blocks suitable for ornamental use, when the color is some shade of blue. More recently, large masses of intense blue color have come from Brazil. In Canada, occurrences around Bancroft, Ontario continue to supply pale-blue to medium-blue lapidary material. The coarse grains of massive sodalite (cleavage faces common), streaks of white or pink material, and large size of blocks serve to identify this mineral.

LAZURITE   Plate 43

$Na_8(S, SO_4, Cl)_4(AlSiO_6)$. Main constituent of lapis lazuli, used in jewelery since prehistoric times. Lazurite is closely intergrown with calcite, diopside, etc., and tiny golden grains of included pyrite are characteristic. For names, see azurite (p. 172). There is a very ancient mine, still being worked, in the Kokcha valley, NE. Afghanistan, in contact-metamorphic limestone. Occasionally the rock contains large (to 3 cm.) dodecahedral crystals of lazurite, much prized by mineral collectors. A paler and considerably poorer grade of lapis lazuli is obtained in quantity from a mine in the Andes of Ovalle, Chile. Minor occurrences on Italian Mountain, Colorado, and Ontario Peak, San Gabriel Mountains, and Cascade Canyon, San Bernardino Mountains, San Bernardino Co., California. Lapis lazuli is much favored for rings, as well as for engraving and for use in mosaic work. Good stones can fetch about \$400 a gram. Also found near Lake Baikal (!!) and in Chile.

SCAPOLITE (WERNERITE) SERIES

Complex aluminum-silicates containing alkalies, chlorine, and carbonate + sulphate radicals; the sodium-rich end member is *marialite* and the calcium-rich end member *meionite;* the name *scapolite* (or *wernerite*) is generally used for specimens which have not received careful analysis to determine exact position in the series. The species are typically found in marble skarns, with pyroxenes, amphiboles, apatite, zircon, and sphene. CANADA – abundant in the Grenville marbles in Quebec and Ontario, as at Faraday Township, Hastings Co., and near Eganville, Renfrew Co., Ontario in crystals (!!!), some of large size. USA – similar crystals to the Canadian in St Lawrence Co., New York and farther south in Orange Co., extending over the border into New Jersey where crystals (!!!) were found in the Franklin mine, Sussex Co. MEXICO – large sharp crystals (!!!) with green diopside in marble at Ayoquesco, Oaxaca. BRAZIL – gem crystals of yellow color, Minas Gerais. MADAGASCAR – gem material at Tsarasaotra; sharp, nearly transparent crystals near Betroka (!!). BURMA – gem material in the gem gravels. Well-formed crystals are simple prismatic (tetragonal) with square to octagonal cross-section; color usually white or cream.

Zeolite group

This group contains a number of minerals whose easily distinguishable external appearance disguises a high degree of similarity in chemical content and composition, as well as in crystal structure. The SiO_4 framework structure contains easily removable water molecules in large openings; with care, this water can be evaporated without destroying the basic crystal structure, and indeed, these structures can reabsorb water, other ions, or even gases. This property is taken advantage of in water softeners because the synthetic zeolites used in them absorb 'hard' calcium ions from household water, and release 'soft' sodium ions from within the zeolites to make the water 'soft'. Nearly all zeolites bubble when fused in the blowpipe. flame, hence the name, from the Greek ζein, to boil. Hydrochloric acid decomposes them, leaving a gelatinous residue of silica.

Natrolite	$Na_2Al_2Si_3O_{10}.2H_2O$	ortho-rhombic	pseudo-tetra-gonal
Scolecite	$CaAl_2Si_3O_{10}.3H_2O$	monoclinic	
Heulandite	$CaAl_2Si_7O_{18}.6H_2O$	monoclinic	
Stilbite	$CaAl_2Si_7O_{18}.7H_2O$	monoclinic	
Harmotome	$BaAl_2Si_6O_{16}.6H_2O$	monoclinic	
Chabazite	$(Ca, Na_2)Al_2Si_4O_{12}.6H_2O$	hexagonal (trigonal)	

NATROLITE-SCOLECITE   Plate 48

STREAK white   H 5–5$\frac{1}{2}$ SG 2.3

COLOR colorless to white; also yellowish, greenish, reddish.

LUSTER vitreous; silky in compact masses.

PROPERTIES cleavage distinct parallel to rectangular prisms, fracture uneven, brittle; transparent to translucent.

CRYSTALS square cross-section prismatic, usually very long, and capped with low pyramid; commonly forms radiating balls or spheres; also finely fibrous, compact.

IDENTIFICATION natrolite and scolecite can scarcely be distinguished except by advanced techniques; the radiating habit and delicate acicular crystals are good clues; only confusion with pectolite, whose crystals are extremely thin, almost fibrous. Natrolite fuses in candle flame while scolecite does not.

The name 'natrolite' means 'sodium-stone'; 'scolecite' comes from the Greek *skolex,* a worm, because it curls up when heated. Natrolite and scolecite form a solid solution series.

HEULANDITE

STREAK colorless   H 4   SG 2.2
COLOR white, colorless, yellowish, faint pink, red, or pale reddish-brown.
LUSTER vitreous except on cleavage surfaces strong pearly.
PROPERTIES perfect cleavage; fracture uneven; brittle; translucent to transparent.
CRYSTALS characteristically coffin-shaped, sometimes diamond-shaped, with much curving; also in coarse granular masses.
IDENTIFICATION shape of crystals and very strong pearly luster; flakes turn white in blowpipe flame and then puff up.

Named after English mineralogist H. Heuland.

STILBITE   Plate 48

CHEMISTRY always appreciable Na.
STREAK white   H $3\frac{1}{2}$-4   SG 2.2
COLOR colorless, white, yellow, grey, reddish, orange, brown.
LUSTER vitreous, cleavage surfaces pearly.
PROPERTIES perfect cleavage but seen only in traces on many specimens because formed of multiple crystals; uneven fracture; brittle; translucent to transparent.
CRYSTALS seldom in good crystals, tending to form flat 'sheaves' of many crystals, curving outwards toward the pointed ends, and some aggregates which spread outward like cauliflowers; also granular massive and in crusts with rounded surfaces.
IDENTIFICATION sheaf-like habit; decomposes in hydrochloric acid but does not form gelatin.

Named after Greek *stilbein,* to 'glimmer', in allusion to luster.
    Harmotome and the similar *phillipsite* form stubby crystals of pseudo-hexagonal outline due to twinning; both are relatively rare and seldom form large crystals.

CHABAZITE

STREAK white · · · H 4½ · · · SG 2.1
COLOR colorless, white with reddish or brownish tinge.
LUSTER vitreous.
PROPERTIES cleavage sometimes distinct, fracture uneven.
CRYSTALS small but common, almost cubic rhombohedrons.
    Interpenetrant twins *(phacolite)*; rarely encrusted or massive.
IDENTIFICATION rhombohedral crystals with poor cleavage.

Named from *chabazios,* one of the stones mentioned in the
    Greek poem *Peri lithon.*
    Formation and occurrence. Zeolites are typical late-hydro-
thermal formations, often found encrusting other minerals. The
widest distribution of this group is in cavities of basic igneous
rocks, in the characteristic paragenesis: zeolites, apophyllite,
analcime, quartz, calcite. As late minerals, zeolites are also
found in ore veins, pegmatite druses, and, sometimes quite
widely, in alpine fissures.
    USA – zeolites found in great abundance in cavities in pillow
basalts of Watchung Mountains of northern New Jersey, as at
Prospect Park and New Street, Paterson, Passaic Co., heulandite
(!!!), stilbite (!!!), chabazite (!!), analcime (!!), with laumontite,
gmelinite, natrolite (!!!), scolecite, mesolite, and thomsonite;
additional species are prehnite (!!!), datolite (!!), quartz (partly
amethyst), calcite, apophyllite (!!!); also various zeolites near
Cascadia, Linn Co., Edwards, Tillamook Co., Ritter Hot
Springs, Lane Co., Oregon and Table Mountain, Jefferson Co.,
Colorado. Very large and partly clear natrolite prisms (!!!) to
2 cm. thick and 6 cm. long from serpentine-chert cavities along
Clear Creek, San Benito Co., California. CANADA – abundant
zeolites in basalts of Nova Scotia, especially along north shores
at Wasson's Bluff, Cape Blomidon, etc., the chabazite of
reddish hue being large and fine (!!!); huge natrolite crystals to
1 m. long reported from serpentine at Thetford Mines, Quebec.
BRAZIL – recently, magnificent specimens of apophyllite,
heulandite, stilbite in cavities in basalt near Veronopolis, Rio
Grande do Sul, the stilbites reaching 10 cm. ICELAND – celebrated zeolite localities, especially at Berufjord, fine stilbite (!!),
heulandite (!!), etc. INDIA – basalts of the Deccan near Poona
provide a large quantity of magnificent specimens as large (to

15 cm. diameter) rosettes of white stilbite, apophyllite crystals (!!!), heulandite, etc.

GERMANY -- Hohenwiel, Swabia, natrolite radiating and as balls in phonolite; Rauschermühle near Kaiserslautern, Rhineland-Palatinate, natrolite with prehnite, analcime, datolite, calcite; Idar-Oberstein, Rhineland-Palatinate, chabazite and harmotome in agate amygdales; St Andreasberg in the Harz Mts., in ore veins (fine pink apophyllite !). AUSTRIA – Obersulzbachtal and Habachtal, Hohe Tauern, zeolites in alpine fissures, e.g. the 'prehnite island' at the Seekar gap, chabazite, stilbite, heulandite, prehnite, smoky quartz, fluorite, apatite, adularia, apophyllite. ITALY – Baveno, chabazite, stilbite, in granite fissures; Val di Fassa, red heulandite, stilbite, natrolite; Siusi Alps and Tiso near Chiusa, chabazite, stilbite, analcime, apophyllite, datolite, rock crystal, amethyst, etc. SWITZERLAND – around the Giuv valley, Uri and Graubünden, e.g. Schattig Wichel, principally stilbite and chabazite; Kleiner Mutsch, classical source of zeolites; Maggia and Ticino valleys, e.g. Alpe Sponda, Cresciano, Biasca, zeolites; Gibelbach near Fiesch, Valais, heulandite, stilbite, chabazite with fluorite etc. USSR – Khibin Mts., natrolite crystals in nepheline-syenite.

APOPHYLLITE

CHEMISTRY $KCa_4F(Si_4O_{10})_2 \cdot H_2O$, potassium-calcium silicate with fluorine and water.

STREAK colorless   TETRAGONAL   H $4\frac{1}{2}$–5   SG 2.3

LUSTER vitreous, but on cleavage surfaces strongly pearly.

PROPERTIES perfect, easy cleavage on base; uneven fracture; brittle; translucent to transparent.

CRYSTALS pseudo-cubes with striated sides; sometimes steeply pointed octahedrons; also granular massive.

IDENTIFICATION crystal shape, cleavage with its pearly luster, striated sides; associated with zeolites.

Named after the Greek *apo-*, 'off', and phyllon, 'leaf', because crystals easily flake off when heated in flame.

Formation and occurrence. Like zeolites.

DATOLITE

CHEMISTRY $Ca(OH)(BSiO_4)$, calcium hydroxyl borosilicate.

STREAK colorless   MONOCLINIC   H 5–$5\frac{1}{2}$ SG 2.9

COLOR faintly greenish, yellowish-green; white massive or colored by inclusions red, yellow, brown, orange, etc.

LUSTER vitreous, somewhat oily on fracture surfaces.

PROPERTIES no cleavage, conchoidal to uneven fracture; brittle; transparent to translucent.

CRYSTALS wedge-shaped with multitude of smaller faces, usually forming druses; also coarse granular and fine-grained massive in nodules.

IDENTIFICATION typical greenish color and shape of crystals; usually associated with zeolites.

The name is after the Greek *dateisthai*, 'to divide', because granular aggregates crumble easily. Large crystals are sometimes cut into gems (Westfield, Massachusetts).

Formation and occurrence. Late-hydrothermal mineral often found coating other species, especially in cavities in basalts and diabases; also in copper deposits of Michigan and in some ore veins; rarely in granites, serpentines. USA – large clear crystals (!!!) Lane Quarry, near Westfield, Massachusetts; fine crystals (!!) Rocky Hill, near Hartford, and Roncari quarry, near East Granby, Connecticut; abundant in quarries of northern New Jersey as druses (!!) with other zeolites at Prospect Park, Passaic Co., especially; curious white, red, orange, porcelain-like nodules to 30 cm. diameter in copper deposits of Keweenaw Peninsula, Michigan, cuttable into cabochon gems and attractive. MEXICO – crystals (!) at Guanajuato. GERMANY – crystals (!!) at St Andreasberg, Harz and in volcanic rocks around Idar-Oberstein, Rhineland-Palatinate. ITALY – crystals (!!), Seiser Alp, Trentino.

ORGANIC COMPOUNDS

A few organic compounds should also be included in the mineral kingdom, such as mellite (tetragonal, honey-yellow crystals, $Al_2C_{12}O_{12}18H_2O$, in lignite) or paraffins such as fichtelite ($C_{19}H_{34}$), found in moorland among the Fichtelgebirge. Coal, petroleum, paraffin waxes and so forth used to be discussed in books on mineralogy, but strictly speaking they belong to geology. Mention should, however, be made of amber, which, though it is actually fossil resin, is a mixture of organic compounds.

Part Three DETERMINATIVE TABLES

How to identify a mineral by its external characteristics is explained in Ch. 3 (p. 55); properties, such as streak, hardness, luster, etc., are also defined there. Ch. 1 discusses crystal forms, and the formation of minerals in general, Ch. 2 their origin.

Page numbers in the last column refer to the description in Part Two.

To recapitulate and summarize the identification 'drill':

1 Decide whether the specimen has non-metallic or metallic luster. NON-METALLIC, pp. 256–89; METALLIC, p. 288 to end of tables.

2 Test the stone for streak by drawing it across a piece of white unglazed porcelain. NON-METALLIC white streak pp. 256–77; yellow to brown p. 276; green p. 280; blue p. 282; red p. 282; grey to black p. 284. METALLIC white streak p. 288; yellow to brown p. 288; red p. 290; grey to black p. 292.

3 Test hardness.

4 By now you should have got it down to two or three possibilities. Check on the page noted in the last column, for any other distinguishing properties.

HARDNESS numbers 1–10 refer to the Mohs scale (p. 56).

TRANSPARENCY

1 = transparent   2 = translucent   3 = opaque
Parentheses ( ) indicate less common condition.

CLEAVAGE

| − = absent | 2 = good | 3+ = highly perfect |
| 1 = poor | 3 = perfect | |

FREQUENCY

* indicates minerals occurring very widely or in large accessible deposits

Hardness	Color: common / others	Luster	Transparency	Streak	Cleavage Quality	Cleavage Form	Fracture	SG	Usual habit and peculiarities
1	colorless / white, light green, yellow, brown	greasy, resinous, dull	2 to 3	white	3+	thin flakes	massive: uneven	2.8	crystals very rare foliaceous to compact; feel: greasy
1 to 2	colorless / red, white yellowish	vitreous	(1) 3	white	—	—	conchoidal	1.6	crystals rare; compact and granular
1 to 2½	green from almost colorless to black	vitreous, pearly, dull	(2) 3	white to green	3+	flaky	splintery	2.5 to 2.9	crystals not common; scaly, foliaceous to massive earthy, often encrusting
2	colorless / white, yellowish	vitreous	1 to 2	white	3	cubic	brittle	2.0	crystals rare; granular masses; soluble in water
2	colorless / white, grey, yellow, blue, red	vitreous	1 to 3	white	3	cubic	conchoidal, brittle	2.1	crystals not rare compact, granular fibrous; soluble water
2	yellow / honey yellow to brown	adamantine, greasy	2 to 3	white	—	—	conchoidal, very brittle	2	crystals fairly rare compact, granular earthy, massive encrustations
2	colorless / white grey, yellow, red, brown	vitreous, pearly	1 to 3	white	3+	rhombohedral flakes	conchoidal fibrous	2.3	crystals and coarse splinters common, compact, granular to massive, fibrous scaly
2	colorless / blue, grey	vitreous	2	white becoming blue	3	flaky	fibrous, smooth	2.6	crystals rare; fibrous, radiating, reniform and earthy
2 to 2½	green from almost colorless to black	vitreous, pearly, dull	(2) 3	white to green	3+	flaky	splintery	2.5 to 2.9	crystals not common; scaly, foliaceous to massive earthy, often encrusting

ystal m	Occurrence (according to formation)	Frequency	Associated with	Name	Refs.
xagonal kes	weakly metamorphic rocks; by decomposition of silicates, also of dolomite	★	actinolite, diopside, olivine, quartz, dolomite, magnesite, etc.	**Talc**	p. 232
	sedimentary like halite		minerals of the saline deposits	**Carnallite**	p. 107
xagonal lets and sms	weakly metamorphic rocks; alteration product of other minerals; alpine fissures	★	magnetite, mica, garnet, actinolite, diopside, feldspar, adularia, quartz	**Chlorite**	p. 237 pl. 13 16 45
bes	sedimentary like halite		halite, minerals of the saline deposits	**Sylvite**	p. 107
bes, rarely tahedra	sedimentary from solutions by the evaporation of water; often mixed with clays	★	gypsum, anhydrite, bitterns	**Halite, Rock Salt**	p. 107 pl. 10 fig. 10
ombic ramids	result of volcanic activity, organic in sediments, decomposition product of sulphides	★	gypsum, anhydrite, celestine, calcite, aragonite	**Sulphur**	p. 72 pl. 2 28
g flaky stals, twins, 21	sedimentary, mainly from alteration of anhydrite and sulphides	★	calcite, dolomite, anhydrite, minerals of the saline deposits, sulphides, sulphur, limonite	**Gypsum**	p. 177 pl. 28
smatic	phosphate pegmatite rocks, secondary formation in iron minerals, peat		phosphate, iron minerals	**Vivianite**	p. 190
xagonal lets and sms	weakly metamorphic rocks; alteration product of other minerals; alpine fissures	★	magnetite, mica, garnet, actinolite, diopside, feldspars, adularia, quartz	**Chlorite**	p. 237 pl. 13 16 45

Hardness	Color: common / others	Luster	Transparency	Streak	Cleavage Quality	Cleavage Form	Fracture	SG	Usual habit and peculiarities
2 to 2½	white / to pale yellow	dull	3	white	—	—	uneven	3.7	no crystals; compact, earthy, massive; encrustations
2 to 2½	colorless / yellowish, greenish, reddish, silvery	pearly, metallic, silky	1 to 3	white	3+	very thin flakes	elastic	2.8	good crystals rare; tabular, flaky, fine scales, massive
2 to 2½	colorless / pale violet, silver-grey, yellowish brown	pearly, metallic sheen	1 to 3	white	3+	very thin flakes	elastic	3.0	no good crystals; fan-shaped or conical to irregular flakes
2½ to 3	dark grey, green, brown, black	pearly, weakly metallic	2 to 3	white	3+	very thin flakes	elastic	3	crystals not frequent; tablets, flakes, foliaceous
2½ to 3	colorless / white	vitreous	1 to 3	white	3	flaky	—	2.4	crystals rare; radiating, fibrous concretionary, encrusting
3	white, yellowish grey, reddish	vitreous	2	white	3	in one direction	brittle	2.1	crystals rare; sugary massive soluble in water
3	colorless / white, many pale colors	vitreous	1 to 3	white	3	rhombohedral	brittle	2.7	crystals very common; granular, massive, earthy, fibrous, bladed
3	colorless / grey, yellowish	adamantine	1 to 2	white	2	in two directions	conchoidal	6.3	crystals common; encrusting
3	colorless / wax, honey and lemon yellow, grey	adamantine	2 to 1	white, yellowish	2	in two directions	conchoidal, brittle	6.8	crystals common; drusy, cavity filling massive

Crystal form	Occurrence (according to formation)	Frequency	Associated with	Name	Refs.
	oxidation zones		as smithsonite	**Hydrozincite**	
hexagonal tablets and prisms	many magmatic and metamorphic rocks, pegmatites, alpine fissures, fragmental in sediments; not in lava deposits	★	mainly rock-forming minerals	**Muscovite**	p. 233 pl. 41
	pneumatolytic rocks		quartz, pneumatolytic cassiterite assemblage	**Zinnwaldite**	p. 236
hexagonal tablets	many magmatic and metamorphic rocks, rarely clastic	★	mainly rock-forming minerals	**Biotite**	p. 235
hexagonal tablets	constituent of bauxite, in cavities, also crystals, in weakly metamorphic rocks		magnetite, hematite, corundam, serpentine, talc, limonite	**Gibbsite Hydrargillite**	p. 152
tabular	sedimentary like halite		minerals of the saline deposits	**Kainite**	p. 107
scalenohedra, flat rhombohedra, hexagonal prisms, twins	sedimentary, stalactitic, hydrothermal alpine fissures, etc.	★	sulphides, quartz, barite, fluorite, dolomite, and many others	**Calcite**	p. 159 pl. 22 23 fig. 20
prismatic, tabular, pyramidal	oxidation zones		galena, limonite, cerussite	**Anglesite**	p. 177
square tablets, flattened pyramids	oxidation zones		galena, sphalerite, siderite, calcite, hydrozincite	**Wulfenite**	p. 183 pl. 29

Hardness	Color: common / others	Luster	Transparency	Streak	Cleavage Quality	Cleavage Form	Fracture	SG	Usual habit and peculiarities
3	yellow, brown, orange	adamantine	2 to 3	white, yellowish	—	—	uneven, splintery	7	crystals common; concretionary, fibrous
3 to 3½	white / red, yellow, grey	greasy, dull	3	white, reddish	3	in one direction	uneven	2.8	no crystals; comp lamellar masses
3 to 3½	colorless / white, grey, yellow, brown, black	adamantine	1 to 2	white, grey	1	—	conchoidal, brittle	6.5	crystals common; compact, reniform tufted, earthy
3 to 3½	colorless / white, blue	vitreous	1 to 3	white	2 to 3	rhombic tablets	conchoidal	4	crystals usually small; coarse, granular, massive, fibrous
3 to 3½	colorless / white, yellowish, black	vitreous	1 to 3	white	3	rhombic tablets	conchoidal	4.5	crystals common; massive, coarsely to finely granular, coarsely flaky
3½	colorless / greenish, yellowish, grey	vitreous	2	white	2	in two directions	conchoidal, brittle	3.7	crystals rare; acicular, bladed, prismatic, radiating
3½	colorless / white, grey, yellowish	vitreous dull	2	white	1	—	uneven, brittle	4.3	crystals rare; botryoidal, radiating, fibrous
3½	colorless / white, yellowish	vitreous	1 to 3	white	—	—	—	2.6	crystals rare; finely granular; soluble in water
3 to 4	colorless / white, grey, bluish, reddish	vitreous	1 to 3	white	2 to 3	cubic	splintery, brittle	2.9	crystals rare; massive, granular, coarsely fibrous
3 to 4	green, white, yellow-brown, yellow, red-brown	dull to sub-resinous	(2) 3	white	—	—	uneven very tenacious	2.6	massive

...stal ...m	Occurrence (according to formation)	Frequency	Associated with	Name	Refs.
...xagonal ...isms, pointed ...ramids	oxidation zones of lead veins		galena, descloizite, pyromorphite, wulfenite, hydrozincite	**Vanadinite**	p. 190
	sedimentary, like halite		minerals of the saline deposits	**Polyhalite**	p. 108
...xagonal ...ramids, ...ttened ...ystals, twins	oxidation zones		galena, limonite, anglesite	**Cerussite**	p. 168
...ismatic and ...ular rhom-... crystals	usually cavity linings in limestone, rarely hydrothermal		calcite, sulphur, gypsum, aragonite	**Celestite**	p. 175 pl. 26
...ombic ...lets, prisms	hydrothermal, sedimentary in concretions, etc.	*	sulphides, manganese minerals, fluorite, calcite	**Barite**	p. 176 pl. 5 27
...xagonal ...isms	hydrothermal, sedimentary		calcite, barite galena	**Strontianite**	p. 168
...xagonal ...ramids	hydrothermal, contact metasomatic		galena, barite	**Witherite**	p. 168
...ramids	sedimentary, like halite		minerals of the saline deposits	**Kieserite**	p. 108
...bic, pris-...tic, tabular	sedimentary, rarely hydrothermal	*	gypsum, calcite, minerals of the saline deposits	**Anhydrite**	p. 173
	hydrothermal alteration of magnesium-rich rocks (olivine, bronzite)	*	magnesite, chromite, olivine, talc, chalcedony, bronzite, etc.	**Serpentine**	p. 238 pl. 42

Hardness	Color: common / others	Luster	Transparency	Streak	Cleavage Quality	Cleavage Form	Fracture	SG	Usual habit and peculiarities
3½ to 4	colorless / white dirty, yellow, brown, grey, black	vitreous	1 to 3	white	3	rhombo-hedral	conchoidal, brittle	2.9	crystals common massive to sacch oidal, friable
3½ to 4	white, grey, brown	vitreous	3	white	3	rhombo-hedral	conchoidal, brittle	2.9 to 3.8	crystals rare; massive to granu
3½ to 4	colorless / white, grey, yellowish, reddish, bluish	vitreous	1 to 2	white	1	in one direction	conchoidal, brittle	2.9	crystals common radiating, acicula fibrous, tuberose nodular, massive
3½ to 4	colorless / green, orange, yellow, brown	adamantine	(2) 3	white	—	—	uneven, brittle	6.5 to 7	crystals common concretionary, reniform, encrus
3½ to 4	colorless / white, grey, yellow, brown	vitreous	1 to 2	white	3	flakes	uneven	2.2	crystals common radiating, sheaf-like aggregates
3½ to 4	colorless / white, grey, yellow to brick-red	vitreous, pearly	1 to 3	white	3	flaky	uneven, brittle	2.2	crystals not rare; flaky and fibrous
4	colorless / all to black	vitreous	1 to 3	white	3	octa-hedral	splintery, brittle	3.2	crystals very com mon; compact, coarsely granula to massive, coars acicular
4	rose-red, raspberry-red	vitreous	2 to 1	white, reddish-white	2	rhombo-hedral	conchoidal	3.5	small crystals; botryoidal, radia ing, sparry
4	colorless / white, yellow, blue	vitreous	1 to 2	white	3	in one direction	—	3.6	good crystals rar radiating, acicula and bladed

stal m	Occurrence (according to formation)	Frequency	Associated with	Name	Refs.
ombohedral, dle-shaped	hydrothermal, by alteration of limestones, sedimentary, in talc veins	★	hydrothermal ores, carbonates, gypsum, anhydrite, talc	**Dolomite**	p. 162 pl. 24
ombohedral	hydrothermal, by alteration of limestones	★	dolomite, siderite, calcite	**Ankerite**	p. 162
isms, pseudo- xagonal ins	hydrothermal, hot springs, organic, sedimentary, product of weathering		clay, gypsum, sulphur, zeolite	**Aragonite**	p. 167 pl. 24
xagonal isms, barrel- aped	oxidation zones of lead veins		galena, cerussite, limonite, vana- dinite, quartz	**Pyromorphite**	p. 189 pl. 30
ombic isms	filling in basalts and pegma- tites, alpine fissures		zeolites, calcite, pegmatite minerals, minerals of alpine fissures	**Stilbite**	p. 251 pl. 48
onoclinic blets	filling in basalts, alpine fissures		zeolites, calcite, minerals of the alpine fissures	**Heulandite**	p. 251 pl. 48
bes, rarely ctahedra	pegmatitic to pneumatolytic, above all hydrothermal, sedimentary	★	galena, sphalerite, other sulphides, quartz, barite, uranium minerals	**Fluorite**	p. 108 pl. 11 20
ombohedra	hydrothermal, sedimentary		sulphides, limonite, hematite	**Rhodochrosite**	p. 156
laded and bular crys- ls, striated	crystalline schists		staurolite, garnet, corundum, muscovite, hornblende	**Kyanite**	p. 200 pl. 33

Hardness	Color: common / others	Luster	Transparency	Streak	Cleavage Quality	Cleavage Form	Fracture	SG	Usual habit and peculiarities
4 to 4½	white, yellow, brown, grey, black, multi-coloured tarnish	vitreous	2 to 3	white, brown, black	3	rhombo-hedral	conch-oidal, brittle	3.8	crystals common coarse to granul radiating, botry-oidal, massive; often with limon coating
4 to 4½	colorless ——— white, brown, black	vitreous	2 to 3	white	3	rhombo-hedral	conch-oidal, brittle	3	crystals common fibrous, granular massive
4½	colorless ——— white, reddish, brownish	vitreous	1 to 2	white	1	—	uneven	2.1	crystals common encrustations
4½ to 5	grey-white, yellowish, brownish	greasy to adamantine	2	white	2	—	conch-oidal, brittle	6	almost always in small crystals or grains, rarely compact
4½ to 5	white, yellowish, reddish	vitreous, silky	2	white	3	in two direc-tions	finely fibrous	2.8	crystals very rare wide or narrow blades, fibrous, encrusting
4½ to 5	colorless ——— white reddish, yellowish, brown	vitreous, pearly	1 to 2	white	3	basal	uneven, brittle	2.3	crystals common; flaky, foliated, granular
5	colorless ——— yellowish, brown, grey, greenish	vitreous	2 to 3	white	2	rhombo-hedral	— brittle	4.4	crystals rare reniform, honeycombed, stalactitic, earthy
5	colorless ——— white, all colors	vitreous	1 to 3	white	1	—	conch-oidal, brittle	3.2	crystals common; massive, granular, fibrous, mamillate concretionary, earthy

ystal m	Occurrence (according to formation)	Frequency	Associated with	Name	Refs.
ombohedra, ten with rved surfaces	hydrothermal, metasomatic, alpine fissures, sedimentary	★	sulphides, calcite, dolomite, ankerite, minerals of alpine fissures	**Siderite**	p. 155 pl. 24 5 6
ombohedra, ten lenticular ystals	metasomatic, metamorphic, sedimentary alteration product	★	chlorite, talc, serpentine, dolomite	**Magnesite**	p. 153
ombohedra	filling cavities in basalt rocks		zeolites, calcite	**Chabazite**	p. 251
pyramidal	pegmatites, pneumatolytic, hydrothermal		pneumatolytic cassiterite assemblage, fluorite, barite, ankerite	**Scheelite**	p. 182
	product of contact-metamorphism of limestones and dolomites		garnet, idocrase, epidote, diopside	**Wollastonite**	p. 231
tragonal isms, ramids and blets	cavity filling in lava deposits and granite, hydrothermal		analcime, zeolites, calcite, sulphides	**Apophyllite**	p. 253
ombohedra	oxidation zones, metasomatic		sphalerite, hydrozincite, hemimorphite, galena, limonite, calcite, dolomite	**Smithsonite**	p. 154
exagonal blets, and isms with ramids	in magmatic rocks, with magnetite ores, pegmatites, alpine fissures, sedimentary	★	magnetite, pegmatite minerals, minerals of alpine fissures, calcite, and many others	**Apatite**	p. 187 pl. 30

Hardness	Color: common / others	Luster	Transparency	Streak	Cleavage Quality	Cleavage Form	Fracture	SG	Usual habit and peculiarities
5	colorless / white, yellow, grey, pale blue	vitreous	2	white	3	prismatic	uneven	3.4	crystals common reniform to stalactitic, fibrous, earthy encrusting
5 to 5½	yellow, green, brownish, sometimes black	adamantine	1 to 3	white	1	—	conchoidal, brittle	3.5	mainly intergrown crystals; rarely granular
5 to 5½	colorless / white, grey, yellow, blood-red	vitreous, often dull	1 to 3	white	—	—	uneven	2.2	crystals common granular to massive, earthy
5 to 5½	colorless / white, grey, yellowish, brownish, reddish	vitreous, silky	1 to 2	white	2	right-angled prisms	conchoidal	2.3	crystals fairly common; mostly acicular, finely fibrous to hair-like, tufts, radiating
5 to 5½	greenish, / white, colorless, yellowish	vitreous	1 to 2	white	—	—	uneven	2.9	crystals common; granular, massive fibrous
5 to 6	blue	vitreous	3	white	1	—	uneven	3	crystals rare; compact
5 to 6	white / colorless, grey, green, red	vitreous, dull	1 to 3	white	3	prismatic	uneven, brittle	2.6	large crystals; granular, massive
5 to 6	brown, green	vitreous, bronze-like	3	white	2	flaky	—	3.3	crystals rare; usually long flat grains
5½ to 6½	colorless / all colors	vitreous	1 to 3	white	—	—	conchoidal	1.9 to 2.5	no crystals; amorphous; concretionary, stalactitic encrusting, etc.

ystal m	Occurrence (according to formation)	Frequency	Associated with	Name	Refs.
	oxidation zones		hydrozincite, sphalerite, galena, wulfenite, limonite, calcite	**Hemimorphite**	p. 208
ular, wedge-aped, twins	in many magmatic and meta-morphic rocks, alpine fissures		hornblende, feld-spars, minerals of alpine fissures	**Sphene**	p. 205 pl. 35
pezo-dra (Fig. 4D)	hydrothermal in ore deposits, cavity filling in basalt rocks		apophyllite, zeolites, calcite	**Analcime**	p. 240 pl. 43
eudo-ragonal isms	cavity fillings in magmatic rocks and especially basic lavas, alpine fissures		apophyllite, other zeolites, calcite	**Natrolite Scolecite**	p. 250 pl. 48
lyhedral onoclinic ystals	in cracks and cavities in basic lavas		zeolites, magnetite, copper	**Datolite** $CaB[OHSiO_4]$	p. 253
	in marble, quartzite		calcite, pyrite	**Lazurite** (lapis lazuli)	p. 172
ragonal isms	hydrothermal alteration pro-duct of feldspars, contact-metamorphic rocks		sanidine, nepheline, idocrase, garnet, diopside	**Scapolite**	p. 249
	in basic magmatic rocks		serpentine, plagioclase	**Bronzite**	p. 228
	hydrothermal alteration of silicates, e.g. in cavities in basalt rocks, hot springs, and sediments	*	quartz, calcite	**Opal**	p. 134 pl. 18

Hardness	Color: common ——— others	Luster	Transparency	Streak	Cleavage Quality	Cleavage Form	Fracture	SG	Usual habit and peculiarities
5½ to 6	blue ——— white, grey, black, colorless	vitreous	1 to 3	white	2	rhombic dodeca-hedral	conch-oidal	2.4	crystals rare; rounded grains; granular, massive
5½ to 6	blue-black ——— yellow, red-brown colorless	resinous, metallic	2	white	3	pyram-idal	conch-oidal, brittle	3.8	only known in crystals
5½ to 6	light grey ——— white, colorless yellowish, brownish, greenish, reddish	vitreous, greasy	1 to 3	white	—	—	conch-oidal	2.6	crystals rare; granular with waxy lustre
5½ to 6	white ——— colorless, grey, yellowish	vitreous	(1) 2	white	—	—	conch-oidal	2.5	mainly crystals; rarely compact
5½ to 6	light red to brown-red	vitreous	2	white	3	in two direc-tions	uneven	3.5	crystals rare; coarse, granular, massive
5½ to 6	all shades of green, grey, white,	vitreous, silky	2 to 3	white	2	oblique-angled prisms	—	3.1	blade-like crystals common; radiating parallel fibres, sheaf-like
5½ to 6	leek- to grey-green	dull sheen	2	white	—	—	splint-ery	3.1	only massive
5½ to 6½	dark green ——— grey, yellow, light to colorless	vitreous	1 to 3	white	2	right-angled prisms	uneven	3.3	good crystals rare bladed, radiating, granular
6	leek-green dark green	vitreous	2	white	2	right-angled prisms	conch-oidal	3.3	good crystals rare grains

stal n	Occurrence (according to formation)	Frequency	Associated with	Name	Refs.
	nepheline-syenites, volcanic rocks		nepheline, leucite	**Sodalite, Hauyne, Nosean**	p. 248
nted or flat agonal amids	alpine fissures		minerals of the alpine fissures	**Anatase**	p. 139
agonal sms	in magmatic rocks like phonolite, eleolite, etc.		aegirite, feldspars, hornblendes, zeolites, zircon, sphene	**Nephelite**	p. 240
pezohedra g. 4D)	in basic volcanic rocks		sanidine, nepheline, augite; never with quartz	**Leucite**	p. 241 pl. 43
	in sediments and metamorphic rocks, contact-metamorphic rocks, hydrothermal		manganese minerals, marble	**Rhodonite**	p. 227
g prisms h rhombic ss-section	in crystalline schists and contact-metamorphosed rocks	★	feldspars, garnet, muscovite; sericite, epidote, chlorite, calcite	**Actinolite**	p. 229 pl. 40
	in serpentinized gabbros, in crystalline schists		serpentine	**Nephrite**	p. 230
sms with are cross-tion g. 23)	various magmatic and meta-morphic rocks, contact-metamorphosed rocks, alpine fissures		garnet, calcite, epidote, idocrase, chlorite, mica	**Diopside**	p. 224 pl. 39
	in contact-metamorphosed limestones, in eclogite		garnet, spinel, idocrase, biotite	**Fassaite**	p. 225

Hardness	Color: common / others	Luster	Transparency	Streak	Cleavage Quality	Cleavage Form	Fracture	SG	Usual habit and peculiarities
6	grey, greenish, yellowish	vitreous	3	white	3	in one direction	uneven	3.3	crystals rare; m... intergrown, co... striated, twiste... blades
6	colorless — often flecked with green	vitreous	1 to 2	white	3	right-angled or slightly oblique prisms	uneven	2.5	nearly always g... crystals; often coated with ch...
6	white — colorless, yellowish, brownish, light flesh-red, labrador-escent (play of color)	vitreous	1 to 3	white	3	right-angled or slightly oblique prisms	uneven	2.5	crystals commo... compact, often large splinters, granular
6	white — colorless, yellowish	vitreous	1 to 3	white	3	slightly oblique prisms	uneven	2.6	crystals commo... granular
6 to 6½	white — colorless, yellowish, grey, green, red, labrador-escent	vitreous	1 to 3	white	3	—	uneven	2.6 to 2.8	crystals rare; splinters, grains
6 to 6½	light green — white, colorless to yellow-green	vitreous	2	white	1	—	uneven	2.9	crystals commo... botryoidal mass... with radiating structure
6½	white to greenish	dull sheen	(2) 3	white	—	—	uneven, splintery	3.2	only massive, r... grains, very tou...
6½	brown, yellow, green, black, blue, pink	vitreous	(1) 2 to 3	white	1	—	uneven	3.4	crystals commo... striated column... aggregates, gra...

stal n	Occurrence (according to formation)	Frequency	Associated with	Name	Refs.
	as epidote, also in amphibolites and eclogite		hornblendes, augite, garnet, epidote, muscovite	**Zoisite**	p. 208
Fig. 24	alpine fissures	★	minerals of the alpine fissures	**Adularia**	p. 245 pl. 45
Fig. 24	in acid magmatic and related rocks, less frequently in metamorphic rocks, clastic in sandstones	★	quartz, mica, plagioclase, tourmaline and many other minerals	**Orthoclase, Microcline, Sanidine**	p. 245 pl. 44 45 47
Fig. 25	in magmatic and metamorphic rocks, pegmatites, alpine fissures	★	orthoclase, quartz, mica, minerals of alpine fissures	**Albite**	p. 246 pl. 46
Fig. 8; ellar nning	important constituent of magmatic and metamorphic rocks, also clastic in sandstones	★	rock-forming minerals, especially hornblendes and augite	**Plagioclase**	p. 246 pl. 46
mbic, millary lets	fissures and cavities of basic volcanic rocks, alpine fissures		zeolites, epidote, calcite, quartz, adularia	**Prehnite**	p. 212 pl. 46
	in crystalline schists, as boulders		serpentine bodies	**Jadeite**	p. 227
ragonal sms and amids	mainly contact-metamorphosed rocks, alteration product, alpine fissures		garnet, diopside, chlorite, talc, mica, calcite	**Idocrase,** (vesuvianite)	p. 211 pl. 36

Hardness	Color: common / others	Luster	Transparency	Streak	Cleavage Quality	Cleavage Form	Fracture	SG	Usual habit and peculiarities
6 to 7	brown to black / colorless, yellow, grey	vitreous, dull	2 to 3	white, weak yellowish, brownish	1	prismatic	conchoidal	7	crystals commo compact, grains fibrous reniforr masses, rolled g
6 to 7	/ white, yellow, colorless	vitreous	1 to 2	white	3	in one direction	—	3.6	good crystals ra radiating, acicul and bladed
6½ to 7	yellowish-grey, grey-green	vitreous, silky	1 to 2	white	3	flat blades	uneven	3.2	crystals very ra acicular to fibro
6½ to 7	olive-green to greenish-yellow, brown / colorless	vitreous	1 to 2	white	1	—	conchoidal	3 to 4	crystals not con mon; grains, granular masses
6½ to 7	brown, grey, blue, greenish	vitreous	1 to 2	white	1	—	conchoidal, brittle	3.3	large crystals; fl needles, lamella
6½ to 7	greenish-white, grey, lilac, pink, yellow, green	vitreous	1 to 3	white	2	right-angled prisms	uneven	3.2	often very large crystals; splinter coarse needles fibrous aggrega
6½ to 7	colorless / grey, yellowish, brown, bluish	vitreous	1 to 2	white	3	flaky	conchoidal, very brittle	3.4	crystals rare; ne like to fibrous,
6½ to 7½	all colors / colorless, mainly brownish-red, never blue	vitreous	(1) to 3	white	—	—	splintery, brittle	4	crystals very con mon; grains, gra lar to massive, rounded fragme

ystal rm	Occurrence (according to formation)	Frequency	Associated with	Name	Refs.
tragonal isms, twins	pegmatitic-pneumatolytic rocks, hydrothermal, placer deposits		pneumatolytic cassiterite assemblage	**Cassiterite**	p. 139 pl. 19
aded and bular crystals, rizontally iated	crystalline schists		staurolite, garnet, corundum, muscovite, hornblende	**Kyanite**	p. 200 pl. 23
	in crystalline schists and veins therein; placer deposits		cordierite, muscovite, garnet, minerals of placer deposits	**Sillimanite**	p. 199
ort prismatic ombic ystals	constituent of basic magmatic rocks, also metamorphic rocks		augite, spinel, magnetite, bronzite, diallage, serpentine	**Olivine**	p. 193 pl. 33
ick triclinic blets with arp edges	mineral of contact-metamorphism, alpine fissures, also in pegmatitic cavities, in granite		calcite, magnetite, minerals of alpine fissures and pegma-	**Axinite**	p. 213
ular, ongated tagonal isms	pegmatites		pegmatite minerals	**Spodumene**	p. 226 pl. 39
ombic blets	in crystalline schists, constituent of many bauxites		hydrargillite, limonite, corundum, serpentine, kyanite	**Diaspore**	p. 152
ombic decahedra, apezohedra	mainly in crystalline schists and contact-metamorphosed rocks, sometimes in magmatic and pegmatitic rocks, in placer deposits	*	muscovite, calcite, idocrase, serpentine, feldspar, hornblendes, etc.	**Garnet family**	p. 194 pl. 31 32 40

Hardness	Color: common / others	Luster	Transparency	Streak	Cleavage Quality	Cleavage Form	Fracture	SG	Usual habit and peculiarities
7	colorless / all colors	vitreous, greasy	1 to 3	white	—	—	conchoidal	2.65	crystals very common; massive, granular, compact acicular, fibrous, botryoidal
7 to 7½	reddish to blackish-brown	vitreous, dull	2 to 3	white	2	in one direction	uneven	3.7	nearly always crystals; often acicular occasionally gran
7 to 7½	grey-blue / colorless, blue, yellow, brown	vitreous	2 to 3	white	1	—	conchoidal	2.6	crystals rare; most grains and massive granular
7 to 7½	black / brown, blue, green, red, colorless	vitreous, pitch-like	1 to 3	white	—	—	uneven	3.1	crystals fairly rare needles, radiating
7½	brownish / colorless	vitreous, dull	(1) to 3	white	1	—	uneven	3.1	crystals fairly rare acicular, radiating crystals often coat with muscovite
7½	brown / colorless, reddish, yellow, grey, green	adamantine, greasy	1 to 3	white	1	—	conchoidal	4.5	nearly always crystals, mostly small grains, never massive
7½ to 8	green / white to yellow, brown, colorless, blue	vitreous	1 to 3	white	1	—	uneven	2.7	nearly always crystals; pebbles
8	colorless / yellow, blue, green, pink	vitreous	1 to 2	white	3	basal	conchoidal	3.5	crystals common; acicular, grains, rarely massive

Crystal form	Occurrence (according to formation)	Frequency	Associated with	Name	Refs.
hexagonal prisms and pyramids (figs. 17, 19)	veins, not constituent of basic magmatic rocks	*	almost all important minerals; not: leucite, some augite, olivine	**Quartz**	p. 127 pl. 14 to 18, 6, 19, 34, 47
prisms, twins crossing each other at various angles	in crystalline schists		muscovite, kyanite, garnet	**Staurolite**	p. 202 pl. 33
	in gneisses, also in magmatic rocks		gravels, quartz, feldspar, garnet, sillimanite, spinel	**Cordierite** (iolite)	p. 217
three-sided prisms	in acid plutonic rocks, also in crystalline schists, sediments and sandstone, in pegmatite	*	quartz, mica, feldspars, pneumatolytic cassiterite assemblage	**Tourmaline**	p. 217 pl. 38 41
square prisms	in crystalline schists, especially in contact zones, in gravels		muscovite, quartz, sillimanite, garnet	**Andalusite**	p. 199 pl. 33
tetragonal prisms with pyramids	in magmatic rocks, and crystalline schists		garnet, diopside, sphene, chlorite, minerals of the gem gravels	**Zircon**	p. 197 pl. 35
hexagonal prisms	pegmatites, hydrothermal in mica schists, in gravels		pegmatite minerals, muscovite, gem gravels	**Beryl**	p. 213 pl. 37
rhombic prisms, often faceted	pneumatolytic, in gravels		cassiterite, quartz, tourmaline, minerals of the gem gravels	**Topaz**	p. 201 pl. 34

Hardness	Color: common / others	Luster	Transparency	Streak	Quality	Form	Fracture	SG	Usual habit and peculiarities
8	red, blue, green, etc. ⎯⎯⎯ colorless,	vitreous	1 (2, 3)	white	1	octahedral	conchoidal	3.7	crystals; grains
9	grey ⎯⎯⎯ cloudy, colorless, blue-grey, yellow-grey, blue, red, green, violet	vitreous	1 to 3	white	2	basal, rhombohedral	conchoidal	4	crystals common; compact, fibrous, granular, grains
10	brown ⎯⎯⎯ yellow, colorless, blue, green	adamantine	3	—	3+	octahedral, brittle	conchoidal	3.52	crystals common; rounded crystalline aggregates.

Non-metallic luster

Hardness	Color: common / others	Luster	Transparency	Streak	Quality	Form	Fracture	SG	Usual habit and peculiarities
1	black	dull	3	dark brown	—	—	friable	c. 1	no crystals; earthy, loosely aggregated, porous
1½ to 2	yellow, brownish	adamantine pearly	2 to 3	yellow	3	flaky	—	3.4	crystals rare; foliated, earthy, powdery
2	yellow ⎯⎯⎯ honey-yellow to brown	adamantine, greasy	2 to 3	white to yellow	—	—	conchoidal, very brittle	2	crystals fairly rare; compact, granular, earthy, massive, encrustations
2	green to sulphur-yellow	vitreous	1 to 3	yellowish	3	flakes	—	3.1	crystals mostly small; scales, u.v. light
2½	honey to lemon-yellow	silky	3	pale yellow	—	—	—	3.8	no crystals; acicular to powdery, u.v. light
2½ to 3	brown, yellow, orange	adamantine	2 to 3	yellowish	—	—	uneven, splintery	7	crystals common; granular massive
3	honey-yellow ⎯⎯⎯ colorless, wax-, lemon-yellow, grey	adamantine	2 to 1	yellowish	2	in two directions	conchoidal, brittle	6.8	crystals common; drusy, cavity filling, massive

...ystal ...m	Occurrence (according to formation)	Frequency	Associated with	Name	Refs.
...ahedra, ...ins (Fig. 7)	contact-metamorphous in limestones and dolomites; in gravels		calcite, dolomite, minerals of the gem gravels	**Spinel**	p. 111
...xagonal ...kes or ...sms, often ...rrel-shaped, ...ellar ...inning	in plutonic rocks and crystalline schists, in gravels	★	magnetite, spinel, muscovite, calcite, dolomite, minerals of the gem gravels	**Corundum**	p. 116 pl. 12
...ahedra, ...ombic ...decahedra, ...ely tetra-...dra	in igneous basic or ultrabasic rocks (volcanic pipes); alluvial deposits		pyrope garnet, chrome diopside, ilmenite, perovskite, zircon	**Diamond**	p. 69 pl. 1

Yellow to brown streak

	Occurrence (according to formation)	Frequency	Associated with	Name	Refs.
	oxidation zones, sediments		manganese oxides, limonite, siderite	**Wad**	p. 144
	hydrothermal (low temperature), hot springs, in clays		realgar, stibnite arsenopyrite, cinnabar, opal	**Orpiment**	p. 106
...ombic ...ramids	result of volcanic activity, organic in sediments, decomposition product of sulphides	★	gypsum, anhydrite, celestite, calcite, aragonite	**Sulphur**	p. 72 pl. 2 28
...uare tabular ...stals	by alteration of pitchblende; in granite fissures, in pegmatites		various uranium minerals, pegmatite minerals	**Autunite**	p. 149
	like autunite, also in fluorite veins		various uranium minerals, fluorite	**Uranophane**	p. 149 pl. 20
...xagonal ...sms, ...ramids	oxidation zones of lead veins		galena, descloizite, pyromorphite, wulfenite, hydrozincite	**Vanadinite**	p. 190
...uare tablets, ...ttened ...ramids	oxidation zones		galena, sphalerite, siderite, calcite, hydrozincite	**Wulfenite**	p. 183 pl. 29

Hardness	Color: common ––– others	Luster	Transparency	Streak	Cleavage Quality	Cleavage Form	Fracture	SG	Usual habit and peculiarities
3½ to 4	brown to black / bright yellow colorless	adamantine to metallic	3 (2, 1)	light yellow to light brown	3	rhombic dodeca-hedral	— brittle	4	crystals common coarse to fine gr... lar masses, botry... oidal, radiating
4 to 4½	brown, yellow, white, grey, black, multi-coloured tarnish	vitreous	2 to 3	brown, black, white	3	rhombo-hedral	conch-oidal, brittle	3.8	crystals common coarse to granul... radiating, botry-oidal, massive; o... coated with lim... nite
4 to 6	black, brown, grey	pitch-like; crystals metallic	3	brown to black	—	—	conch-oidal	9 to 10.5	crystals rare; massive, compac... reniform
5 (to 1)	brown to black, yellow ochre	adamantine to metallic, dull	3	brown	3	in one direction	splintery	4.3	crystals rare; fibr... massive, compac... earthy, loose, po...
5 to 5½	black to dark brown	adamantine to metallic	3 (2)	brown to yellow	3	in one direction	uneven	7.3	good crystals no... common, someti... large; coarsely radiating to flak...
5½	iron-black	dull	3	brown	—	—	conch-oidal, brittle	4 to 4.8	crystals very rare... rounded grains, compact masses
5½ to 6	green-black to black	vitreous, horny	3	brown	2	oblique-angled prisms	splintery	3.2	crystals not com-mon; acicular, fibrous, granular
5½ to 6	yellow-brown to red-brown	adamantine to metallic	1 to 2	yellow to brown	1	—	conch-oidal	4	only known as crystals
5½ to 6	brown, pitch-black to green, sometimes red sheen	vitreous, sometimes partly metallic	3	yellow to brown	2	flaky	—	3.5	good crystals rar... mainly granular ... foliaceous

ystal m	Occurrence (according to formation)	Frequency	Associated with	Name	Refs.
rahedra to ombic decahedra	hydrothermal, skarn minerals, sedimentary	★	galena, other sulphides, quartz, calcite, fluorite,	**Sphalerite**	p. 78 pl. 3
ombohedra, en with rved surfaces	hydrothermal metasomatic alpine fissures sedimentary	★	sulphides, calcite, dolomite, ankerite, minerals of alpine fissures	**Siderite**	p. 155 pl. 24 5 6
bes	pegmatite, hydrothermal, sedimentary		uranophane and torbernite, Co-Ni-Ag-sulphides, fluorite, barite	**Pitchblende Uraninite**	p. 146 pl. 20
ismatic edles	oxidation zones, alteration product of iron minerals	★	iron minerals, sulphides, manganese oxides, calcite and many others	crystals (H5) **Goethite** others (H1–5) **Limonite**	p. 151 pl. 21
ck tabular to ismatic rstals	pegmatitic, pneumatolytic, hydrothermal		quartz, scheelite, pneumatolytic cassiterite assemblage	**Wolframite**	p. 145 pl. 29
tahedra	liquid-magmatic rocks, gravels		serpentine, basic-rock minerals	**Chromite**	p. 114
-sided isms ig. 23)	many magmatic and metamorphous rocks, contact-metamorphosed rocks, magnetite deposits	★	feldspars, garnet, mica, chlorite	**Hornblende**	p. 230 pl. 40
kes	alpine fissures		minerals of alpine fissures	**Brookite**	p. 139
ombic prisms th pyramids	in basic magmatic rocks; also in cordierite gneisses		plagioclase, olivine, hornblende, magnetite, sphene	**Hypersthene**	p. 228

Hardness	Color: common / others	Luster	Transparency	Streak	Cleavage Quality	Cleavage Form	Fracture	SG	Usual habit and peculiarities
6	iron-black to brownish	pitchlike, dull	3	brown, cherry-red, black	1	in one direction	conchoidal	5 to 8	crystals fairly cor mon; compact masses
6 to 6½	greenish to black, reddish	vitreous	3	yellow to dark green	2	right-angled prisms	uneven	3.5	crystals often lar also fibrous and hairlike
6 to 6½	yellow-brown / red, brown-red, black	adamantine to metallic	2 to 3	yellow-brown	2	prismatic	conchoidal	4.2	crystals common grains, acicular, needles in rock crystal
6 to 7	brown to black / colorless, yellow, grey	vitreous, dull	2 to 3	yellowish to brownish	1	prismatic	conchoidal	7	crystals common compact, grains, fibrous reniform masses, rolled gra

Non-metallic luster

1 to 2½	green from almost colorless to black	vitreous, pearly, dull	(2) 3	green, white	3+	flaky	splintery	2.5 to 2.9	crystals not com-mon; scaly, folia-ceous to massive, earthy, often en-crusting
2 to 2½	emerald- to grass-green	vitreous	2	pale green	3+	flaky	—	3.5	crystals often goo but small; scaly
3½ to 4	bright to dark green	silky, vitreous	(2) 3	bright green	1	—	conchoidal, brittle	4	crystals very rare; acicular, radiating botryoidal, diver-gently fibrous, ear
5½ to 6	green-black to black	vitreous, horny	3	grey-green	2	oblique-angled prisms	splintery	3.2	crystals not com-mon; acicular, fibrous, granular
5½ to 6	grey-green to brown-black, bronze	pearly, metallic	3	grey-green	3	flaky	uneven	3.3	no good crystals; foliaceous to lamellar masses
5½ to 6	black-green to black	vitreous	2	grey-green	2	right-angled prisms	uneven	3.5	good crystals not common; acicular and lamellar masse

...stal ...m	Occurrence (according to formation)	Frequency	Associated with	Name	Refs.
...ck tabular ...ystals with ...isms	only in granitic pegmatites		pegmatite minerals	**Columbite**	p. 146
...ort and long ...uare to octa-...dral prisms	syenitic and associated rocks		nepheline, horn-blendes, feldspars	**Aegirite**	p. 227
...ragonal ...isms often ...ngitudinally ...iated, twins	magmatic and metamorphic rocks, pegmatites, alpine fissures, placer deposits	*	ilmenite, magnetite, apatite, minerals of alpine fissures	**Rutile**	p. 138 pl. 19
...ragonal ...isms, knee-...aped twins	pegmatitic-pneumatolytic rocks, hydrothermal, placer deposits		pneumatolytic cassiterite assem-blage	**Cassiterite**	p. 139 pl. 19

Green streak

...stal ...m	Occurrence (according to formation)	Frequency	Associated with	Name	Refs.
...sided ...lets and ...sms	weakly metamorphic rocks; alteration product of other minerals; alpine fissures	*	magnetite, mica, garnet, actinolite, diopside, feldspar, adularia, quartz	**Chlorite**	p. 237 pl. 13 16 45
...all tablets	by alteration of pitchblende		uranium minerals, quartz, limonite, hematite	**Torbernite**	p. 149
	oxidation zones of copper deposits		azurite, limonite, copper minerals	**Malachite**	p. 172 pl. 25
...sided ...sms ...g. 23)	many magmatic and meta-morphous rocks, contact-metamorphosed rocks, magnetite deposits	*	feldspars, garnet, mica, chlorite	**Hornblende**	p. 230 pl. 40
	basic plutonic rocks and serpentine		plagioclase, serpentine	**Diallage**	p. 222
...uare prisms	skarn minerals		magnetite, garnet, zinc blende, calcite	**Hedenbergite**	p. 224

Hardness	Color: common / others	Luster	Transparency	Streak	Cleavage Quality	Cleavage Form	Fracture	SG	Usual habit and peculiarities
5½ to 6	black to green	vitreous, pitchlike, dull	3	grey-green	2	right-angled prisms	uneven	3.4	crystals fairly common; grains, granular, radiating
6 to 6½	greenish to black, reddish	vitreous	3	yellow to dark green	2	right-angled prisms	uneven	3.5	crystals often large, also fibrous and hairlike

Non-metallic luster

Hardness	Color: common / others	Luster	Transparency	Streak	Cleavage Quality	Cleavage Form	Fracture	SG	Usual habit and peculiarities
2	blue, grey / colorless	vitreous	2	white, becoming blue	3	flaky	fibrous, smooth	2.6	crystals rare; fibrous radiating, reniform and earthy
3½ to 4	azure blue, bright to dark blue	vitreous	2	sky-blue	2	in one direction	uneven, brittle	3.8	crystals common; radiating, earthy

Non-metallic luster

Hardness	Color: common / others	Luster	Transparency	Streak	Cleavage Quality	Cleavage Form	Fracture	SG	Usual habit and peculiarities
1½	dark red	adamantine, greasy	2	orange	1	—	conchoidal	3.5	crystals rare; massive, granular, encrusting
2 to 2½	scarlet to cinnabar red	adamantine	2	light yellow to red	2	rhombohedral	splintery, brittle	5.6	crystals rare; massive in veins, encrusting
2 to 2½	red / black-red, grey, brown	crystals adamantine, otherwise metallic to dull	3 (2)	red	2	rhombohedral	uneven, brittle	8.1	crystals rare; massive, compact, earthy, encrusting
3 to 3½	white / red, yellow, grey	greasy, dull	3	reddish-white	3	in one direction	uneven	2.8	no crystals; compact fibrous prismatic aggregates
3 to 4	steel grey	oily to metallic, dull	3	reddish-brown, black	—	—	conchoidal, brittle	4.4 to 5.4	crystals fairly common; compact masses
3½ to 4	red-brown, cochineal red, grey	adamantine, dull	1 to 3	red to brown-red	2	octahedral	conchoidal, brittle	6	crystals common; granular to massive, hair-like; very often coated with malachite

ystal m	Occurrence (according to formation)	Frequency	Associated with	Name	Refs.
agonal oss-section, ins	many basic magmatic rocks, especially volcanic rocks, in crystalline schists and contact-metamorphic rocks	★	plagioclase, garnet	**Augite**	p. 225 pl. 39
ort and long uare to octa-dral prisms	syenitic and associated rocks		nepheline, hornblendes, feldspars	**Aegirite**	p. 227

Blue streak

ystal m	Occurrence	Frequency	Associated with	Name	Refs.
smatic	phosphate pegmatite rocks, secondary formation in iron minerals, peat		phosphate, iron minerals	**Vivianite**	p. 190
ort, flat smatic stals	oxidation zones of copper deposits		malachite, copper minerals	**Azurite**	p. 171 pl. 25

Red streak

ystal m	Occurrence	Frequency	Associated with	Name	Refs.
ort longi-dinally iated prisms	hydrothermal (low-temperature) hot springs, as volcanic sublimate		orpiment, cinnabar, opal, stibnite	**Realgar**	p. 105 pl. 9
ismatic	hydrothermal		silver minerals, galena, calcite	**Proustite**	p. 104
rahedra	low temperature dissemina-tions and impregnations		stibnite, pyrite, arsenopyrite, chalcedony	**Cinnabar**	p. 89 pl. 7
	sedimentary like halite		minerals of the saline deposits	**Polyhalite**	p. 108
rahedra and ombic decahedra g. 15	hydrothermal		sulphides, quartz, calcite, barytes	**Tetrahedrite**	p. 83 pl. 5
tahedra	oxidation zones		malachite, azurite, limonite, copper	**Cuprite**	p. 110

Hardness	Color: common / others	Luster	Transparency	Streak	Cleavage Quality	Cleavage Form	Fracture	SG	Usual habit and peculiarities
4	pink to red	vitreous	2 to 1	reddish-white, white	2	rhombo-hedral	conchoidal	3.5	crystals; globular, radiating fibrous, massive
6	iron black to brownish	pitchlike, dull	3	cherry-red, brown, black	1	in one direction	conchoidal	5 to 8	crystals fairly common; compact masses
6½	black, light to dark red, red-brown	dull, looks polished	3	red, light red	—	—	—	5.1	no crystals; compact to massive, earthy oolitic, radiating. reniform, fibrous
6½	brown-red	vitreous	2	cherry-red	2	in one direction	uneven	3.4	crystals rare; small needles, radiating granular

Non-metallic luster

Hardness	Color: common / others	Luster	Transparency	Streak	Cleavage Quality	Cleavage Form	Fracture	SG	Usual habit and peculiarities
1½	blue-black	dull	3 (2)	shiny black	3	flaky	even	4.7	crystals very rare coarse, powdery, earthy, as coating and encrustations
2	lead grey to iron black	dull (tarnishes)	3	shiny grey	1	—	conchoidal, can be cut with a knife	7.3	crystals rare; compact, reticulated, flaky, powdery, encrusting
2½ to 3	dark lead-grey	dull (tarnishes)	3	shiny grey	1	—	conchoidal, can be cut with a knife	5.6	crystals rare and small; compact, earthy, powdery, encrusting
3 to 3½	white / colorless, grey, yellow, brown, black	adamantine, dull	3 (1 to 2)	grey-white	1	—	conchoidal, brittle	6.5	crystals common massive; earthy
3 to 4	steel-grey	oily to metallic, dull	3	black, reddish-brown	—	—	conchoidal, brittle	4.4 to 5.4	crystals fairly common; compact masses
5 to 5½	dark brown to black	adamantine to metallic	3 (2)	black, brown, yellow	3	in one direction	uneven	7.3	good crystals not common, sometimes large; coarsely radiating to flaky

stal n	Occurrence (according to formation)	Frequency	Associated with	Name	Refs.
mbohedra	hydrothermal, sedimentary		sulphides, limonite, hematite, manganese oxides	**Rhodochrosite**	p. 156
k tabular stals with ms	only in granitic pegmatites		pegmatite minerals	**Columbite**	p. 146
	hydrothermal, sedimentary	*	minerals of iron-ore deposits, sediments	**Hematite**	p. 118 pl. 13
	manganese deposits in crystalline schists		manganese minerals	**Piemontite**	p. 207

Grey to black streak

agonal es	alteration product of copper sulphides		all copper sulphides	**Covellite**	p. 90
es	hydrothermal, oxidation-cementation zones		galena, silver, silver minerals, cerussite	**Argentite**	p. 77
es	hydrothermal veins and impregnations, cementation zones		bornite, covelline, other copper minerals, pyrite	**Chalcocite**	p. 75
matic stals, twins	oxidation zones		galena, limonite, anglesite	**Cerussite**	p. 168
ahedra and mbic ecahedra . 15)	hydrothermal		sulphides, quartz, calcite, barite	**Tetrahedrite**	p. 83 pl. 5
k tabular rismatic tals	pegmatitic, pneumatolytic, hydrothermal		quartz, scheelite, pneumatolytic cassiterite assemblage	**Wolframite**	p. 145 pl. 29

Hardness	Color: common / others	Luster	Transparency	Streak	Cleavage Quality	Cleavage Form	Fracture	SG	Usual habit and peculiarities
5 to 6	black, brownish black, grey-black	dull, looks polished	3	brown-black	—	—	—	4.5	no crystals; com earthy, fibrous, stalactitic
5 to 6	iron-black	dull, looks polished	3	black, brownish	—	—	uneven	4.7	crystals fairly ra massive, granul
5½ to 6	black to brown and green	vitreous, metallic	3	black	2	in one direction	uneven	4.1	good crystals ra coarsely bladed;
5½ to 6	black	dull	3	black, grey-black	1	octahedral	conchoidal, brittle	5	crystals common granular to mass magnetic
5½ to 6	green-black to black	vitreous, horny	3	grey-green	2	oblique-angled prisms	splintery	3.2	crystals not com mon; acicular, fibrous, granular
5½ to 6	grey-green to brown-black, bronze	pearly, metallic	3	grey-green	3	flaky	uneven	3.3	no good crystals foliaceous to lamellar masses
5½ to 6	black to black-green	vitreous	2	grey-green	2	right-angled prisms	uneven	3.5	good crystals not common; acicula and lamellar mass
5½ to 6	black to green	vitreous, pitchlike, dull	3	grey-green	2	right-angled prisms	uneven	3.4	crystals fairly com mon; grains, gran lar, radiating
5½ to 6	pitch-black, brown to green, sometimes red sheen	vitreous, sometimes partly metallic	3	grey, yellow to brown	2	flaky	—	3.5	good crystals rare mainly granular to foliaceous
6	iron-black to brownish	pitchlike, dull	3	black, brown, cherry-red	1	in one direction	conchoidal	5 to 8	crystals fairly com mon; compact masses
6 to 7	dark green to yellow-green, sometimes grey, black	vitreous	2	grey	2	in one direction	uneven	3.4	crystals fairly com mon; small needle radiating, massive

tal	Occurrence (according to formation)	Frequency	Associated with	Name	Refs.
	sedimentary deposits, oxidation zones barite	★	limonite, calcite	**Psilomelane**	p. 144
ets; mbohedral	magmatic segregations, pegmatites, alpine fissures, detrital deposits	★	magnetite, hematite, garnet, apatite, rutile	**Ilmenite**	p. 122
nbic ms	contact-metamorphosed and pneumatolytic rocks		augite, hornblendes, magnetite	**Ilvaite**	p. 211
hedra, nbic ecahedra, s (Fig. 7)	veins, placer deposits	★	sulphides, hematite, hornblendes, augite, apatite, garnet, chlorite	**Magnetite**	p. 113 pl. 13
sided ms . 23)	many magmatic and metamorphous rocks, contact-metamorphosed rocks, magnetite deposits	★	feldspars, garnet, mica, chlorite	**Hornblende**	p. 230 pl. 40
	basic plutonic rocks and serpentine		plagioclase, serpentine	**Diallage**	p. 222
are prisms	skarn-minerals		magnetite, garnet, zinc blende, calcite	**Hedenbergite**	p. 224
gonal s-section, ns	many basic magmatic rocks, especially volcanic rocks, in crystalline schists and contact-metamorphic rocks	★	plagioclase, garnet	**Augite**	p. 225
mbic prisms pyramids	in basic magmatic rocks; also in cordierite gneisses		plagioclase, olivine, hornblende, magnetite, sphene	**Hypersthene**	p. 228
k tabular tals with ms	only in granitic pegmatites		pegmatite minerals	**Columbite**	p. 146
gated ted crystals	in crystalline schists, alpine fissures, also as a rock-forming mineral		garnet, idocrase, augite, hornblende, chlorite, calcite	**Epidote**	p. 206 pl. 32

Hardness	Color: common / others	Luster	Transparency	Streak	Cleavage Quality	Cleavage Form	Fracture	SG	Usual habit and peculiarities
8	dark green, grey-green, black	vitreous	3	grey	1	octa-hedral	conch-oidal	4.3	intergrown cry granular

Metallic luster

Hardness	Color: common / others	Luster	Transparency	Streak	Cleavage Quality	Cleavage Form	Fracture	SG	Usual habit and peculiarities
2 to 2½	greenish, reddish, silvery ——— colorless	pearly, silvery-metallic, silky	1 to 3	white	3+	very thin flakes	— elastic	2.8	good crystals ra tabular, flaky, fi scales, massive
2½ to 3	black, brown-black	metallic	3	white	3+	very thin flakes	—	3	small flakes and scales
5 to 6	brown, green, bronze	metallic sheen	3	white	2	flaky	—	3.3	good crystals ra mostly long flat grains
5½ to 6	colorless ——— yellow, red-brown, blue-black	metallic, adamantine	2	white	3	pyram-idal	conch-oidal, brittle	3.8	only known as crystals
6 to 7	colorless ——— brown to black, yellow, grey	partly sub-metallic, adamantine	2 to 3	white, pale yellowish, brownish	1	pris-matic	conch-oidal	7	crystals commo compact grains, fibrous reniform masses, rolled g

Metallic luster

Hardness	Color: common / others	Luster	Transparency	Streak	Cleavage Quality	Cleavage Form	Fracture	SG	Usual habit and peculiarities
2½ to 3	golden yellow to silver-white	metallic	3	—	—	—	uneven, mal-leable, sectile	15.5 to 19.5	crystals not com mon; rounded masses, 'nugget grains, scales
3½ to 4	black to dark brown	metallic to adamantine	3	brown	3	rhombic dodeca-hedral	— brittle	4	crystals commo coarse to fine granular masses
5 to 5½	light copper-red, brown tarnish	metallic, dull	3	brownish-black	1	—	uneven	7.7	crystals rare; fin granular to mas reniform

stal 1	Occurrence (according to formation)	Frequency	Associated with	Name	Refs.
hedra	in garnet and cordierite gneisses, contact-metamorphosed rocks		sulphides, garnet, cordierite	**Gahnite**	p. 113

White streak

sided ets and ms	many magmatic and metamorphic rocks, pegmatites, alpine fissures, fragmental in sediments, not in lava deposits	*	mainly rock-forming minerals	**Muscovite**	p. 233 pl. 41
	in decomposed rocks of different kinds, except sediments	*	rock-forming minerals	**Biotite**	p. 235
	in basic magmatic rocks		serpentine, plagioclase	**Bronzite**	p. 228
ted or flat agonal mids	alpine fissures		minerals of alpine fissures	**Anatase**	p. 139
agonal ms, knee- ed twins	pegmatitic-pneumatolytic rocks, hydrothermal, placer deposits	*	pneumatolytic cassiterite assemblage	**Cassiterite**	p. 139 pl. 19

Yellow to brown streak

hedra; s	hydrothermal in veins; gravels		quartz, silver minerals, sulphides, carbonates, fluorite	**Gold**	p. 66 pl. 1
ahedra to mbic ecahedra	hydrothermal, skarn minerals	*	galena, other sulphides, quartz, calcite, fluorite, barite	**Sphalerite**	p. 78 pl. 3
	hydrothermal		nickel and copper sulphides, other sulphides, barite, calcite, siderite	**Niccolite**	p. 87

Hardness	Color: common / others	Luster	Transparency	Streak	Cleavage Quality	Cleavage Form	Fracture	SG	Usual habit and peculiarities
5 to 5½	dark brown to black	metallic to adamantine	3	brown, yellow	3	in one direction	uneven	7.3	good crystals not common, some large; coarsely radiating to flaky
5½	iron-black	sub-metallic	3	brown	—	—	conchoidal, brittle	4 to 4.8	crystals very rare, round grains, compact masses
5½ to 6	pitch-black, brown to green, partly red sheen	metallic, partly vitreous	3	yellow to brown	2	flaky	—	3.5	good crystals rare, usually granular foliaceous
5½ to 6	yellow-brown to red-brown	metallic, adamantine	1 to 2	yellow to brown	1	—	conchoidal	4	only known as crystals
6 to 6½	brown / red, brown-red, yellow, colorless, black	metallic to adamantine	(1) to 3	yellow-green	2	prismatic	conchoidal	4.2	crystals common, grains, acicular, needles in rock crystal

Metallic luster

Hardness	Color: common / others	Luster	Transparency	Streak	Cleavage Quality	Cleavage Form	Fracture	SG	Usual habit and peculiarities
2 to 2½	red / black-red, grey, brown	metallic to dull	3 (2)	red	2	rhombohedral	uneven, brittle	8.1	crystals rare; massive, compact, earthy, encrusting
2 to 2½	dark red to lead-grey, black	metallic, also adamantine	2	cherry-red	2	rhombohedral	splintery, brittle	5.8	crystals not common; compact, intergrown, encrusting
2½ to 3	copper red / dark and multi-coloured tarnish	metallic, dull	3	red	—	—	hackly; ductile and malleable	8.9	distinct crystals plates, thin sheets, grains, nodules, arborescent dendr
3 to 4	steel-grey, faded tarnish	metallic, oily	3	reddish-brown, black	—	—	conchoidal, brittle	4.4 to 5.4	crystals fairly common; compact masses

tal	Occurrence (according to formation)	Frequency	Associated with	Name	Refs.
tabular ismatic als	pegmatitic-pneumatolytic to hydrothermal		quartz, scheelite, pneumatolytic cassiterite assemblage	**Wolframite**	p. 145 pl. 29
hedra	liquid magmatic rocks, placer deposits		serpentine, olivine, bronzite	**Chromite**	p. 114
nbic ns with mids	in basic magmatic rocks, also in cordierite gneisses		plagioclase, olivine, hornblende, magnetite, sphene	**Hypersthene**	p. 228
crystals	alpine fissures		minerals of the alpine fissures	**Brookite**	p. 139
gonal ns often itudinally ted, twins 3)	magmatic and metamorphic rocks, pegmatites, alpine fissures, placer deposits	*	ilmenite, magnetite, minerals of alpine fissures	**Rutile**	p. 138 pl. 19

			Red streak		
hedra	low-temperature disseminations and impregnations		stibnite, pyrites, arsenopyrite, chalcedony	**Cinnabar**	p. 89
natic to nohedral als	hydrothermal		silver minerals, galena, calcite	**Pyrargyrite**	p. 104
s, hedra	oxidation and cementation zones		cuprite, chalcocite, azurite, malachite, calcite	**Native copper**	p. 62 pl. 1
hedra and nbic ecahedra 15)	hydrothermal		sulphides, quartz, calcite, barite	**Tetrahedrite**	p. 83 pl. 5

Hardness	Color: common / others	Luster	Transparency	Streak	Cleavage Quality	Cleavage Form	Fracture	SG	Usual habit and peculiarities
3½ to 4	cochineal-red, red-brown, grey	metallic, adamantine	2 to 3	red to brown-red	2	octa-hedral	conch-oidal, brittle	6	crystals commo… granular to mas… hairlike; often coated with ma…

Metallic luster

1	steel-grey	metallic, dull	3	grey	3+	flaky	—	2.2	crystals very ra… curved, scaly p… acicular, granul… earthy
1½	lead-grey	metallic	3	grey, powdery green	3	flaky	smooth, flexible	4.8	crystals very ra… coarse and scal… flakes
1½	blue-black	submetallic, dull	3 (2)	shiny black	3	flaky	even	4.7	crystals very ra… coarse, powder… earthy, as coati… and encrustatio…
2	lead-grey, dark tarnish	metallic	3	lead-grey	3	flaky	hackly, flexible	4.6	crystals commo… radiating, mass…
2	lead-grey to iron-black	metallic very dark tarnish	3	shiny grey	1	—	conch-oidal, sectile	7.3	crystals rare; co… pact, reticulated… flaky, powdery
2½ to 3	dark lead-grey	metallic, dark tarnish	3	shiny grey	1	—	conch-oidal, sectile	5.6	crystals rare and small; compact,… earthy, powder… as encrustations…
2½ to 3	lead-grey	metallic, tarnishes	3	grey-black	3	cubic	brittle, some-times flexible	7.4	crystals commo… compact, coarse… fine granular, ro… out as 'lead glas…
2 to 3	black, steel grey	metallic	3	black	—	—	very brittle	about 4.5	good crystals ve… rare; coarsely crystalline, com… earthy, acicular
3	reddish-brown, multi-coloured tarnish	metallic	3	grey-black	—	—	conch-oidal	5.1	crystals very rar… compact, flaky

stal m	Occurrence (according to formation)	Frequency	Associated with	Name	Refs.
ahedra	oxidation zones		malachite, azurite, limonite, copper	**Cuprite**	p. 110

Grey to black streak

	in primary metamorphic sedimentary rocks, pegmatites		rock-forming minerals, calcite, pegmatite minerals	**Graphite**	p. 69 pl. 1
	pegmatitic-pneumatolytic rocks, hydrothermal		pyrites and quartz, pneumatolytic cassiterite assemblage	**Molybdenite**	p. 99
agonal es	alteration product of copper sulphides		all copper sulphides	**Covellite**	p. 90
ngated, dle-like stals	hydrothermal gangues and replacement veins		lead, zinc and silver sulphides, cinnabar, quartz	**Stibnite**	p. 91 pl. 7
es	hydrothermal, oxidation and cementation zones		galena, native silver, silver minerals, cerussite	**Argentite**	p. 77
es	hydrothermal veins and impregnations, cementation zones		bornite, covelline, other copper minerals, pyrites	**Chalcocite**	
es, ahedra, yhedra, nded ners (Fig. 3)	mainly hydrothermal, sedimentary	★	zinc blende, other sulphides, quartz, calcite, fluorite, barite, siderite	**Galena**	p. 87 pl. 6
	residual sediments, oxidation zones	★	limonite, barite, calcite	**Pyrolusite**	p. 144 pl. 21
	pegmatites, hydrothermal, sometimes in alpine fissures, sedimentary		chalcopyrite, chalcocite, covellite, pyrites, magnetite	**Bornite**	p. 76

Hardness	Color: common / others	Luster	Transparency	Streak	Cleavage Quality	Cleavage Form	Fracture	SG	Usual habit and peculiarities
3	steel grey to black	metallic, looks polished	3	grey	—	—	conchoidal	5.8	crystals comm[on] granular to ma[ss]
3½	steel grey to iron black	metallic to adamantine	3	grey-black	3	prismatic	brittle	4.4	crystals rare; radiating, fibro[us]
3 to 4	steel grey	metallic, oily	3	black, reddish-brown			conchoidal, brittle	4.4 to 5.4	crystals fairly c[om]mon; compact masses
3½ to 4	brass-yellow, multi-coloured to black tarnish	metallic	3	black	—	—	uneven, brittle	4.2	crystals comm[on] compact, massi[ve] coatings
4	brown-black	metallic	3	black, dark-brown	3	in one direction	uneven, brittle	4.3	crystals comm[on] radiating, acicu[lar]
4	bronze reddish tarnishes	metallic	3	black-grey	—	—	conchoidal, brittle	4.6	crystals very ra[re] compact, coars[e] flakes, massive
5	silver-white, grey, tarnishes	metallic	3	black-grey	1	—	uneven, brittle	7.3	crystals small a[nd] rare; granular, columnar, acicu[lar]
5 to 5½	black to dark brown	metallic to adamantine	3	black, brown	3	in one direction	uneven	7.3	good crystals n[ot] common, some large; coarsely radiating to fla[t]
5 to 5½	light copper red, brown tarnish	metallic, dull	3	black, brownish	1	—	uneven	7.7	crystals very ra[re] finely granular massive, renifo[rm] often coated wi[th] green
5½	silver-white, reddish tarnish	metallic	3	black	1	—	uneven, brittle	6.2	crystals comm[on] compact, granu[lar]
5 to 6	iron black	metallic	3	black to brownish black	—	—	uneven	4.7	crystals fairly c[om]mon; compact, granular, iron r[ose] weakly magneti[c]

stal n	Occurrence (according to formation)	Frequency	Associated with	Name	Refs.
mbic lets, twins	hydrothermal		tetrahedrite, sulphides	**Bournonite**	p. 105
ated rhom- prisms	hydrothermal		chalcocite, pyrites	**Enargite**	p. 83
ahedra and mbic lecahedra g. 15)	hydrothermal		sulphides, quartz, calcite, barite	**Tetrahedrite**	p. 83 pl. 5
embling rahedra	veins, mainly hydrothermal, skarn minerals, placer deposits	★	mainly sulphides, quartz, calcite, barite	**Chalcopyrite**	p. 82 pl. 4
gitudinally ated prisms	sedimentary in oolites and massive, hydrothermal		manganese oxides, calcite, barite	**Manganite**	p. 152
agonal lets	liquid magmatic, hydro- thermal, gravel deposits	★	pyrites, chalcopyrite, arsenopyrite	**Pyrrhotite**	p. 84 pl. 4
	pneumatolytic to hydro- thermal		arsenopyrite, Ni and Co minerals	**Loellingite**	p. 98
ck tabular prismatic stals	pegmatittic-pneumatolytic to hydrothermal		quartz, scheelite, pneumatolytic cassiterite assem- blage	**Wolframite**	p. 145 pl. 29
	hydrothermal		chloanthite, skutterudite, other sulphides, barite, calcite, siderite	**Niccolite**	p. 87
ritohedra g. 16D) bes	skarn minerals, hydrothermal veins		chalcopyrite, pyrrhotine, garnet, copper, nickel and silver minerals	**Cobaltite**	p. 97
lets, mbohedra	magmatic segregations, pegmatites, alpine fissures, placer deposits	★	magnetite, hema- tite, garnet, apatite, rutile	**Ilmenite**	p. 122

Hardness	Color: common ----- others	Luster	Transparency	Streak	Cleavage		Fracture	SG	Usual habit and peculiarities
					Quality	Form			
5½ to 6	grey-green to brown-black, bronze	metallic sheen	3	grey-green	3	flaky	uneven	3.3	no good crystals foliaceous to lamellar masses
5½ to 6	pitch black brown to green, sometimes red sheen	metallic, vitreous	3	grey, brown	2	flaky	—	3.5	good crystals rar mainly granular foliaceous
5½ to 6	silver-white to steel grey	metallic	3	black	2	parallel to prism faces	uneven, brittle	6	crystals common compact, radiatir finely granular
5½ to 6	black	metallic	3	black, grey-black	1	octa-hedral	conch-oidal, brittle	5	crystals common granular to mass magnetic
5½ to 6	black to brown	metallic to vitreous	3	black	2	in one direc-tion	uneven	4.1	good crystals rar coarsely bladed
5½ to 6	tin white, reddish tarnish	metallic	3	black	1	—	conch-oidal	6.6 to 7.2	crystals small; co pact; frequently coated with coba bloom
5½ to 6	tin white, greenish tarnish	metallic	3	black	—	—	uneven, brittle	6.6 to 7.2	crystals small; granular to massi frequently coated with nickel bloor
6 to 6½	gold-yellow, yellow-brown,	metallic, dull	3	black	—	—	conch-oidal, brittle	5.1	crystals very com mon; coarsely granular to massi radiating, dendrit often coated with limonite
6½	steel grey to iron black	metallic tarnishes	3	black, dark red	3	poor	conch-oidal	5.1	crystals common; compact, folia-ceous, iron roses

ystal m	Occurrence (according to formation)	Frequency	Associated with	Name	Refs.
	basic plutonic rocks and serpentinized rocks		plagioclase, serpentine	**Diallage**	p. 222
ombic isms with ramids	basic magmatic rocks, also cordierite gneisses		plagioclase, olivine, hornblende, magnetite, sphene	**Hypersthene**	p. 228
nple roof-aped crystals, e prism iated	pneumatolytic, skarn minerals, hydrothermal	★	sulphides, pneumatolytic cassiterite assemblage, gold and quartz	**Arsenopyrite**	p. 98
tahedra, ombic decahedra, ins (Fig. 7)	veins, gravels	★	sulphides, hematite, hornblendes, augite, apatite, garnet, chlorite	**Magnetite**	p. 113 pl. 13
	contact metamorphosed and pneumatolytic rocks		augite, hornblendes, magnetite	**Ilvaite**	p. 211
bes	hydrothermal, cobalt, nickel, silver, uranium veins		Co-Ni-Ag minerals, sulphides	**Skutterudite, Safflorite**	p. 103
bes	hydrothermal, cobalt, nickel, silver, uranium veins		Co-Ni-Ag minerals, sulphides	**Chloanthite Rammels-bergite**	p. 103
bes with ooth iations, ritohedra, c. (Fig. 16)	veins	★	other sulphides primarily	**Pyrite**	p. 92 pl. 8
ky, rhombo-dral crystals ith rounded rfaces	veins; sediments	★	nearly all minerals	**Hematite**	p. 118 pl. 13

General bibliography

Selected reading for the interested amateur; a choice of good books on many aspects of mineralogy and related fields.

BATEMAN, A. M. *The Formation of Mineral Deposits*, New York, John Wiley & Sons; London, Chapman & Hall, 1951.

BERRY, L. G. & MASON, B. *Mineralogy*, San Francisco, W. H. Freeman & Co., 1959.

DANA, E. S. *Manual of Mineralogy*, 17th ed. rev. by C. S. Hurlbut, Jr., New York, John Wiley & Sons, 1959. *Minerals and How to Study Them*, 3rd ed. rev. by C. S. Hurlbut, Jr., New York, John Wiley & Sons, 1949.

DIETRICH, R. V. Mineral Tables (Hand Specimen Properties of 1500 Minerals), *Bulletin Virginia Polytechnic Institute*, vol. 59, no. 3, Blacksburg, 1966.

EICHHOLZ, D. E. *Pliny Natural History*, vol. 10 [on stones and gemstones], London, Heinemann Ltd, 1962.

FENTON, C. L. & M. A. *The Rock Book*, Garden City, New York, Doubleday & Co., 1950.

GLEASON, S. *Ultraviolet Guide to Minerals*, New York, Van Nostrand Reinhold Co., 1960.

HOLDEN, A. & SINGER, P. *Crystals and Crystal Growing*, Garden City, New York, Doubleday & Co., 1960.

HURLBUT, C. S., Jr. *Minerals and Man*, New York, Random House; London, Thames & Hudson, 1969.

MARRISON, L. W. *Crystals, Diamonds, and Transistors*, London, Penguin Books, 1966.

METZ, R. *Precious Stones and Other Crystals*, London, Thames & Hudson, 1964.

READ, H. H. (*ed.*) *Rutley's Elements of Mineralogy*, 25th ed., London, Allen & Unwin, 1962.

SINKANKAS, J. *Gemstones of North America*, New York, Van Nostrand Reinhold Co., 1959.

Gem Cutting, 2nd ed., New York, Van Nostrand Reinhold Co., 1962.

Mineralogy for Amateurs, New York, Van Nostrand Reinhold Co., 1964.

Prospecting for Gemstones and Minerals, 2nd ed., New York, Van Nostrand Reinhold Co., *in press*, 1970.

SMITH, G. F. H. *Gemstones*, 13th ed. rev. by F. C. Phillips, London, Methuen & Co. Ltd, 1958.

WEBSTER, R. *Gems, Their Sources, Descriptions and Identification*, 2nd ed., London, Butterworths, 1970.

Magazines and Journals:

The Australian Lapidary Magazine, Jay Kay Publications, 11 Robinson St, Croydon, New South Wales.

Canadian Rockhound, P.O. Box 194, Station A, Vancouver 1, British Columbia.

Earth Science, Box 550, Downers Grove, Illinois 60515.

Gems and Minerals, P.O. Box 687, Mentone, California 92359.

Gems, the British Lapidary Magazine, 29 Ludgate Hill, London, E.C.4.

Lapidary Journal, P.O. Box 2369, San Diego, California 92112.

Index